Cellular Mobile Systems Engineering

The Artech House Mobile Communications Series

John Walker, Series Editor

For a complete listing of *The Artech House Telecommunications Library*, turn to the back of this book.

CELLULAR MOBILE SYSTEMS ENGINEERING

Saleh Faruque

Artech House
Boston • London

Library of Congress Cataloging-in-Publication Data
Faruque, Saleh.
  Cellular mobile systems engineering / Saleh Faruque.
    p.   cm.
  Includes bibliographical references and index.
  ISBN 0-89006-518-7 (alk. paper)
  1. Cellular radio—Design and construction.   I. Title.
  TK6570.M6F37 1996
  621.3845'6—dc20   96-46063
                                                         CIP

British Library Cataloguing in Publication Data
Faruque, Saleh
  Cellular mobile systems engineering
  1. Cellular radio
  I. Title
  621.3'8456

  ISBN   0-89006-518-7

Cover design by Jennifer Makower.

© 1996 ARTECH HOUSE, INC.
685 Canton Street
Norwood, MA 02062

International Standard Book Number: 0-89006-518-7
Library of Congress Catalog Card Number: 96-46063

10 9 8 7 6 5 4 3 2

*To Yasmin, inspiring me by example, and Shams, driving me by curiosity,
I lovingly dedicate this book*

▼▼▼

Contents

▼▼▼

PREFACE

Today, with the advent of radio technology, cellular communication has reached all walks of life, bringing communities and businesses closer than ever before. It enables us to communicate with anyone at any time from anywhere within the service area. Yet its applications in various service environments are still in their infancy. Radio technology itself is still evolving. The switching equipment, which is essentially a multiple access point, requires long-haul wires for intersystem and intrasystem connectivity. Wires have yet to be removed from computer communication systems as well as from residential premises. Needless to say, today's cellular communication systems are largely based on wired links while a fraction of them are wireless. Moreover, the communication system itself lacks a global standard. It is the responsibility of the scientific and the industrial community around the globe to bring this technology to maturity.

In a concerted effort to augment this process, this book presents a comprehensive overview of cellular communication systems currently in use worldwide. Considerable effort has been made to blend theory and practice together. Numerous illustrations are provided for the basic understanding of the subject.

Beginning with a concise description of the characteristics and development of the cellular radio, this book quickly moves on to give in-depth coverage of the latest AMPS, TDMA, and CDMA techniques and suggest new approaches to deployment issues, organized as follows:

Chapter 1: Introduction to Cellular Technology. A brief history of cellular radio. North American dual-mode AMPS-TDMA bands and systems. North American dual-mode AMPS-CDMA bands and systems. North American PCS bands. Salient features of cellular radio.

Chapter 2: Elements of the Cellular Communication System. The generic system. Cell site configuration & communication protocol. The public switching center and its role. The mobile switching center and its role. T1 and DS1 link. Access methods.

Chapter 3: North American Dual-Mode AMPS-TDMA. Advanced Mobile Phone System (AMPS). Specifications of AMPS and TDMA. AMPS/TDMA call processing and hand-off algorithms. AMPS/TDMA coexistence. Voice channel and control channel description and operation. Control channel capacity. Cochannel and adjacent channel interference issues.

Chapter 4: Introduction to CDMA. Spectrum spreading/despreading techniques. PN codes, orthogonal codes, and Walsh codes. Process gain. Soft and hard capacity. CDMA power control. Soft and hard hand-off. Transmit/receive structures.

Chapter 5: Propagation Prediction. The basic concept of RF propagation. Multipath propagation. Propagation modeling. Statistical analysis. Computer-aided propagation analysis. RF survey and optimization.

Chapter 6: The Art of Traffic Engineering. The basic concept. Probability and statistics. Elements of traffic engineering. Use of the Erlang table. Cell site provisioning.

Chapter 7: Frequency Planning. The basic concept. OMNI and sectorization techniques. Interference analysis. Capacity evaluation and enhancement techniques.

Chapter 8: Cell Site Engineering. Site selection and coverage prediction. Base station provision. Base station design. Antenna engineering. Cell site maintenance.

Cellular Mobile Systems Engineering is a fingertip reference for engineers designing and developing cellular systems, managers involved in systems planning, and researchers and instructors of cellular courses. It is also a primer for final-year undergraduate and graduate-level engineering students. The contents of this book are now being offered as standard course material by the Northern Telecom (NORTEL) training department for its customers.

I would like to thank NORTEL for granting me quiet time with my own work.

In closing, I would like to say a few words about how this book was conceived. My wife, Yasmin, a writer and crackerjack grammarian who lives in a world all her own, had an ardent desire to see her own work in print. The tremendous urge that propelled her also motivated me. I thank her graciously for the generous contribution of her time and effort throughout the progress of this book. The lively, inquiring mind of our son, Shams, taking his first steps of discovery into this fascinating world of ours, drove me to put my best into this work.

If this book is of the least assistance to my readers, I shall be amply rewarded.

CHAPTER 1
▼▼▼

INTRODUCTION TO CELLULAR TECHNOLOGY

1.1 HISTORICAL OVERVIEW

The quest to know the unknown and see the unseen is inherent in human nature. It is this restlessness that has propelled mankind to ever higher pinnacles and ever deeper depths. This insatiable desire led to the discovery of light as being electromagnetic, paving the way to the discovery of the radio.

The origin of radio can be traced back to the year 1680 to Newton's theory of composition of white light. He postulated that white light was a composition of various colors. This theory brought the importance as light as an area of study to the attention of many scientists, especially those in Europe, who began to pursue experiments with light. Table 1.1 is a list of important early discoveries connected to the eventual development of the radio. These discoveries are the foundation of today's wireless communication systems.

Apparently, it took many civilizations to develop the radio as a result of experiments with light. Light experiments also resulted in several other discoveries, not directly related to the radio, as listed in Table 1.2.

Experiments with light are still being carried out today in many universities and industries. One of the outcomes of light experiments in the 1970s is the *optical fiber,* which is currently being used for long-haul voice and data transmission. It is

Table 1.1
Early Developments Leading to the Discovery of Radio

1680	Theory of composition of white light	Newton
1800	Infrared ray	Herschel
1801	Ultraviolet ray	Ritter
1805	Wave theory of light	Thomas Young
1805	Proved wave theory; transverse light wave	Fresnel
1831	Relationship between light, electricity, and magnetism	Faraday
1864	Electromagnetic theory, $c = f\lambda$	Maxwell
1886	Proved Maxwell's theory by Lyden jar experiment	Hertz
1897	Radio	Marconi

Table 1.2
Experiments With Light

1880	Fingerprint	Herschel
1895	X-rays	Roentgen
1900	Gamma rays	Rutherford
1960s	Cosmic ray	—

believed that the use of optical fiber technology will increase dramatically with the introduction of wideband networks for voice, data, and video transmission such as the advanced intelligent network (AIN), which is based on the asynchronous transfer mode (ATM) switch.

Figure 1.1 is a composite representation of related discoveries about light that is known as the electromagnetic spectrum, which is arranged according to light frequencies. These frequency components differ in energy according to the wavelength. Their ability to propagate also differs in different propagation media. All electromagnetic waves are transverse in nature.

Notice in Figure 1.1 that radiowaves and microwaves occupy a small fraction of the spectrum, approximately 300 GHz, located at the beginning of the spectrum. This communication band is further divided into several sub-bands as shown in Table 1.3. The cellular band falls within the UHF band (300 to 3000 MHz). Only a fraction of this band is allocated for cellular communication systems.

1.2 PROGRESS IN RADIO COMMUNICATIONS

A good description of the last 100 years of progress in radio communication can be obtained from [1] and [2]. A brief overview of these developments is presented in Tables 1.4 and 1.5.

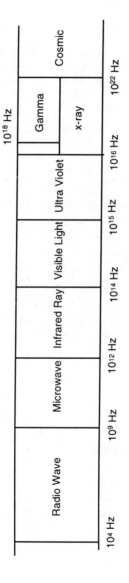

Figure 1.1 Electromagnetic spectrum.

Table 1.3

Communication Sub-Bands

Very low frequency (vLF)	3 to 30 kHz
Low frequency (LF)	30 to 300 kHz
Medium frequency (MF)	300 to 3000 kHz
High frequency (HF)	3 to 30 MHz
Very high frequency (VHF)	30 to 300 MHz
Ultra high frequency (UHF)	300 to 3000 MHz
Super high frequency (SFH)	3 to 30 GHz
Extra high frequency (EHF)	30 to 300 GHz

Table 1.4

Early Radios

1914	Radio first used for practical communication in World War I
1920s	First radio broadcast
1940s	Radar used in World War II
1950s	First commercial television

Table 1.5

Evolution of Public Mobile Telephony

1960s	First mobile telephony, MTS and IMTS
1970s/1980s	AMPS, NAMPS, GSM, DAMPS (TDMA)
1990s	CDMA

1.3 CELLULAR CONCEPT

Today's wireless communication systems are based on a composite wireless and wired system as shown in Figure 1.2 where the wireless segment of the communication system is shown as a cluster of seven hexagonal cells. This is a widely used cell plan, also known as the $N = 7$ frequency plan in which all the available channels are evenly distributed among seven cells.

Each cell is essentially a radio communication center where a mobile subscriber establishes a call with a land telephone through the mobile telephone exchange (MTX) and the public switching telephone network (PSTN). This composite platform enables us to communicate with anyone at any time from anywhere within the service area. Note that MTX and PSTN are essentially multiple access points serving as system intelligence. It is this technology that has reached all walks of life, bringing

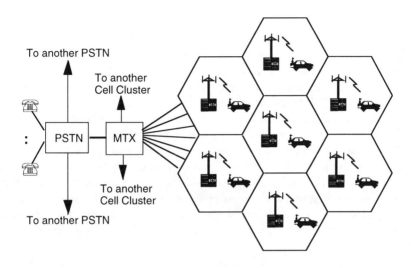

Figure 1.2 Basic elements of a cellular communication system.

communities and businesses closer than ever before. Designing, implementing, and maintaining this complex network is a challenging task that falls under several engineering disciplines such as RF propagation, antenna engineering, frequency planning, traffic engineering, and cell site provisioning. A good compromise between theory and practice is the essence of this communication system.

1.4 CELLULAR FREQUENCY BANDS

Due to the rapid growth of cellular traffic in recent years, it was necessary for the federal regulatory authority to allocate new frequency bands. As a result, rapid technological growth took place. A brief description of these technologies and the supporting frequency bands are given in the following subsections.

1.4.1 North American AMPS and TDMA Bands

In 1974, the Federal Communication Commission (FCC) allocated a 40-MHz band, known as the nonexpanded spectrum (NES), in the 800- to 900-MHz range for cellular application. The AMPS (advanced mobile phone system) standard, introduced in 1979, was subsequently adopted by the FCC. Licenses to operate these frequencies were issued in 1982 for wireline (Band B) and nonwireline (Band A) markets. Because of the rapid growth of cellular traffic, an additional 10-MHz band, known as the expanded spectrum (ES), was allocated.

Cellular telephony provides *full-duplex* communications, which requires simultaneous transmission in both directions: (1) forward path or downlink and (2) reverse path or uplink. A 45-MHz guardband is provided to avoid interference between Tx/Rx channels as shown in Figure 1.3.

The available frequency bands are further divided into several sub-bands. Each band occupies 12.5 MHz, of which 10 MHz is NES and 2.5 MHz is ES. Being a multiple-access technique, these bands are divided into several channels of 30 kHz each, and each channel is assigned to a subscriber. This arrangement provides 10 MHz/30 kHz = 333 channels per band in the NES and 2.5 MHz/30 kHz = 83 channels per band in the ES. The total number of channels per band is therefore 333 + 83 = 416 in each band. Among them, 21 channels are used as control channels, which are used for channel assignment, paging, messaging, etc. The remaining 395 channels are used as voice channels for conversations. Figure 1.4 illustrates the access technique used in AMPS where each FDMA channel is assigned to a singe mobile.

With the advent of AMPS, which is based on frequency-division multiple access (FDMA), rapid technological growth took place. In the 1990s, additional communication standards were proposed for cellular, such as time-division multiple access (TDMA) and code-division multiple access (CDMA).

In the TDMA (IS-54) standard, each AMPS channel is time shared by six mobiles (presently three mobiles) as shown in Figure 1.5. When one mobile has access to the channel, the other two are idle. This is accomplished by a special frame structure. TDMA channel capacity is therefore 3 × FDMA or 6 × FDMA.

1.4.2 North American CDMA Bands

The North American CDMA standard (IS-95) was introduced in 1994. It is based on the spread-spectrum (SS) technique in which multiple users have access to the same band. This is accomplished by assigning a unique orthogonal code (Walsh code) to each user. The spreading is also accomplished by means of the same code. This code is generated from a *pseudonoise* (PN) sequence whose chip rate (switching rate) is many times that of the input binary data. The chip rate of the PN sequence was chosen to be approximately 1.25 MHz to derive 10 different frequency bands from the existing 12.5-MHz cellular carrier (A or B) as shown in Figure 1.6. Each of these bands can support 64 Walsh codes. Various attributes of codes and the associated capacity issues are discussed in Chapter 4.

CDMA air-link is based on a forward link and a reverse link, separated by 45 MHz. This is similar to AMPS and TDMA as shown in Figure 1.7. The forward link is comprised of four different link protocols: (1) a pilot channel, (2) a paging channel, (3) a sync channel, and (4) a traffic channel. The reverse link is based on two different link protocols: (1) an access channel and (2) a traffic channel. It has

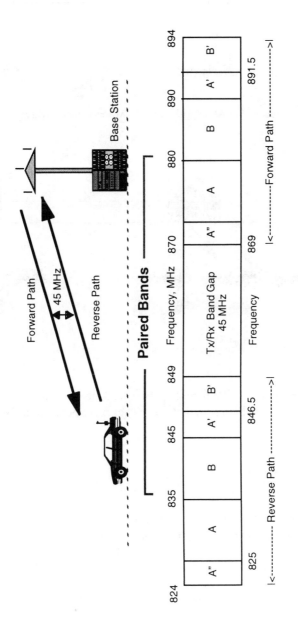

Figure 1.3 Cellular band allocation.

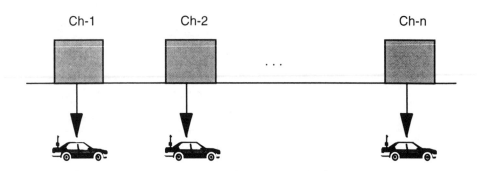

Figure 1.4 Access technique used in AMPS where each FDMA channel is assigned to a single mobile.

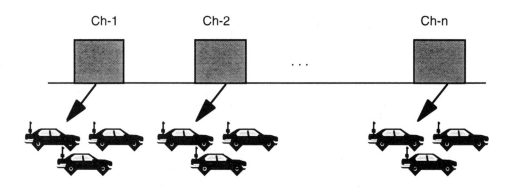

Figure 1.5 TDMA-3. Each AMPS channel is time shared by three mobiles.

been claimed that CDMA cellular will outperform the existing TDMA cellular. At present, CDMA trials are being carried out by several operators in North America.

1.4.3 North American PCS Bands

The North American personal communication services (PCS) standard was introduced in 1994. Six different bands are designated for major trading areas (MTAs) and basic trading areas (BTAs). All MTAs are allocated a 15-MHz band, whereas all BTAs are allocated 5-MHz bands for a total of 60-MHz bands, as shown in Figure 1.8. This is also a full-duplex system having a 20-MHz band gap. This band gap is allocated to unlicensed operators for voice and data at 10 MHz each.

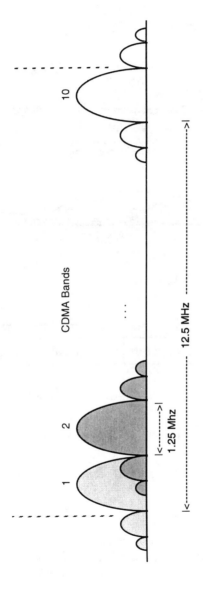

Figure 1.6 CDMA uses ten 1.25-MHz bands from the existing 12.5-MHz cellular carrier (A or B).

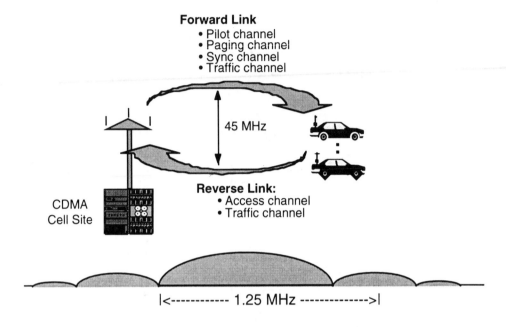

Figure 1.7   CDMA band. Each band is 1.25 MHz wide, supports 64 Walsh codes, and operates in full-duplex mode.

MTA	BTA	MTA	BTA	BTA	BTA	Data	Voice	MTA	BTA	MTA	BTA	BTA	BTA
A 15 MHz	D 5 MHz	B 15 MHz	E 5 MHz	F 5 MHz	C 15 MHz	10 MHz	10 MHz	A 15 MHz	D 5 MHz	B 15 MHz	E 5 MHz	F 5 MHz	C 15 MHz

1850 (MHz)   1910 (MHz)   1930 (MHz)   1990 (MHz)

|<-------- **Lower Band,** 60 MHz, Licensed ---------->|<--Unlicensed -->| <-------- **Upper Band,** 60 MHz, Licensed ------------>|

Figure 1.8   North American PCS bands, operating in full-duplex mode.

1.5   SALIENT FEATURES OF CELLULAR RADIO

1.5.1   Mobility

Mobility is one of the major features of cellular communication systems. It implies that a cellular call, originating from anywhere, any time within the service area, would be able to maintain the same call without service interruption, while in motion. This is attributed to the built-in hand-off mechanism, which is a process of changing the carrier frequency. Its primary purpose is to assign a new frequency while a mobile moves into a new cell as shown in Figure 1.9. This is accomplished by setting

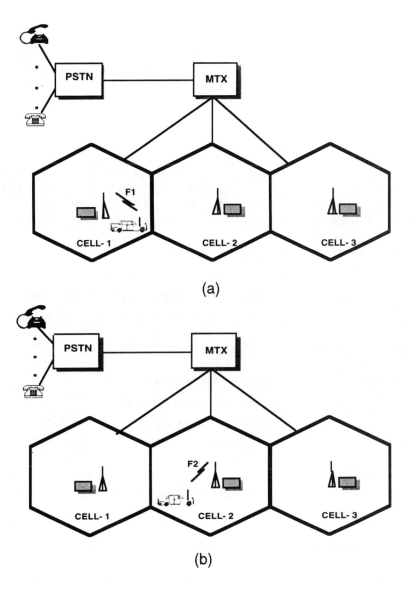

Figure 1.9 Illustration of mobility. (a) A mobile originating a call in cell 1. The MTX assigns a channel and the mobile tunes to the corresponding frequency. (b) The mobile moves into cell 2, its signal is weak, the MTX assigns a new channel from cell 2, the mobile tunes to the new frequency, and the conversation continues without service interruption. This is known as the hand-off process and requires <200 ms.

a hand-off threshold; that is, if the received signal level is too low and reaches a predefined threshold, the system controller—namely the mobile switching center (MTX)—provides a stronger free channel (frequency) from an adjacent cell. This process continues from cell to cell as long as the mobile is in the coverage area, maintains the call, and is also in motion.

1.5.2 Capacity

Channel capacity is measured by the available voice channels per cell, translated into Erlangs. Because there are 21 control channels (see Chapter 7), the number of voice channels is $333 - 21 = 312$ in the NES and $416 - 21 = 395$ in the ES. The channel capacity per cell or per cluster in a seven-cell plan can be evaluated as:

$$\text{NES channel capacity} = 312/7 \approx 44 \text{ voice channels per cell}$$
$$= 35 \text{ Erlangs/cell @ 2\% GOS}$$
$$= 35 \times 7 = 245 \text{ Erlangs/cluster @ 2\% GOS}$$
$$\approx 7350 \text{ subscribers (@ 30 subs/Erlangs typical)}$$

Similarly with ES, the channel capacity becomes

$$\text{ES channel capacity} = 395/7 = 56 \text{ channels per cell}$$
$$= 46 \text{ Erlangs/cell @ 2\% GOS}$$
$$= 46 \times 7 = 322 \text{ Erlangs/cluster @ 2\% GOS}$$
$$\approx 9,660 \text{ subscribers (@ 30 subs/Erlangs typical)}$$

Assuming 10 square miles/cell in a rural area, a cluster of seven cells translates into 7,350 subscribers/7,000 square miles in the NES and 9,660 subscribers/7,000 square miles in the ES.

1.5.3 Frequency Reuse

Because there are a limited number of channels, these channel groups are reused at regular distance intervals. This is an important engineering task that requires a good compromise between capacity and performance.

The mechanism that governs this process is called *frequency planning*. Several frequency planning techniques are available today. One of the most widely used frequency planning technique is the $N = 7$ frequency plan. In this plan, all the available channels are divided into 21 frequency groups, numbered 1 to 21. These frequency groups are then evenly distributed among seven cells, three groups per cell as shown in Figure 1.10. This group (cluster) of 7 cells is then reused along with the associated frequency groups according to Figure 1.10 where the original cluster is surrounded by six additional clusters. A 7-cell cluster is further surrounded by 12 clusters (not shown). In this manner, a given service area can be covered by clusters of cells with frequency reuse, thus providing mobility and capacity.

However, with the advent of subscriber growth, the existing frequency plans do not provide adequate service due to limitations such as cochannel interference and capacity. Therefore, a mechanism is needed to enhance cellular performance and capacity.

In this book, we present a directional frequency reuse plan that yields an additional carrier-to-interference ratio (C/I), determined by the antenna front-to-

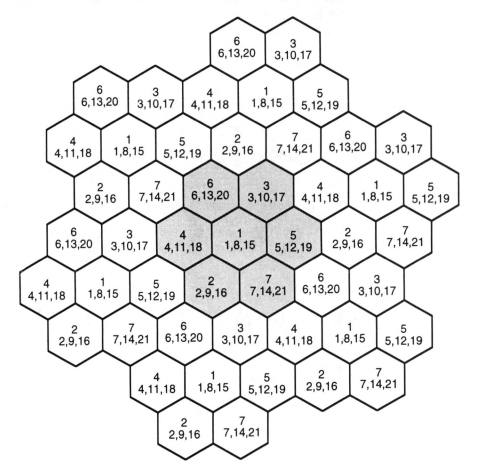

Figure 1.10 The $N = 7$ frequency reuse plan.

back ratio. In this method, a group of channels is reused in the same direction in which the antenna is pointing, thus fully exploiting the antenna front-to-back ratio (Figure 1.11). Because a typical directional antenna has more than a 21-dB front-to-back ratio, a frequency can be reused more often as long as it repeats in the same direction, thereby enhancing AMPS-TDMA capacity (see Chapter 8).

1.5.4 Roaming

There are many cellular operators within the same city, using different switches, radios, and cell site equipment. But a subscriber is registered to one operator only. As a result, an agreement between these operators is needed to provide services to any subscriber irrespective of a call's origin. This is accomplished by means of a separate link between various switches, which is established by the IS-41 link protocol (Figure 1.12). A mobile moving out of its own territory and establishing a call from

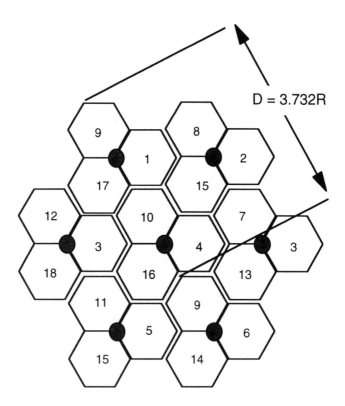

Figure 1.11 Directional frequency reuse plan.

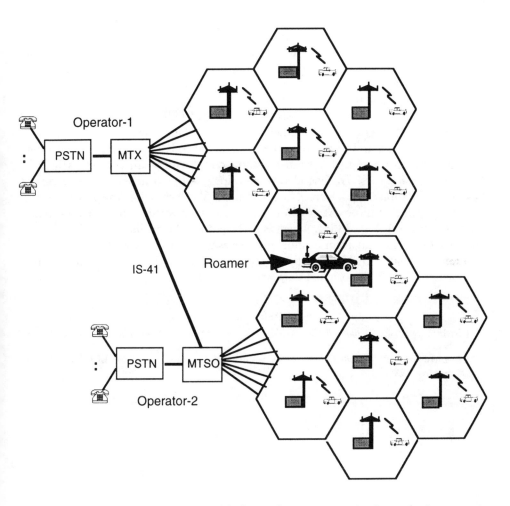

Figure 1.12 Illustration of roaming. A mobile from a host system moving into a foreign system is designated as a roamer. Communication can take place if there is a link (IS-41) between mobile switches.

a foreign territory is known as a *roamer* and the process itself is known as *roaming*. Roaming enhances mobility.

1.5.5 Data Transmission

Data transmission over the analog cellular channel is a new feature in today's cellular communication systems. This is attributed to the recent introduction of cellular digital packet data (CDPD). It is based on the existing cellular networks and uses an unused AMPS channel from a given frequency set in a cell site. Numerous services can be performed and received over the CDPD channel, including fax transmission, computer data transmission, stock exchange portfolio access, news wire service access, commodity price list access, travel information, and many more services that currently rely on e-mail.

The CDPD channel is composed of a forward CDPD (FOCDPD) channel and a reverse CDPD (RECDPD) channel over which data transmission takes place between the base station and the mobile. Data transmission is based on encoding 47 data bits into a (63, 47) Reed-Solomon code, which can correct as high as $(63 - 47)/2 = 8$ errors. The encoded word (block) is then transmitted by means of GMSK modulation at 19.2 Kbps. Note that in GMSK modulation, the power spectral density is such that more than 90% energy is retained within a transmission bandwidth of $19.2 \times 1.2 = 23$ kHz, which is sufficient for the available 30-kHz channel spacing. As a result, adjacent channel interference is greatly reduced and is expected to be lower than TDMA. On the receiving side, the RF signal is demodulated, decoded and finally the original data are recovered.

References

[1] Mehrotram, Asha, *Cellular Radio, Analog and Digital Systems,* Norwood, MA: Artech House, 1994.

[2] Lee, William C. Y., *Mobile Cellular Telecommunications Systems,* New York: McGraw-Hill Book Company, 1989.

CHAPTER 2
▼▼▼

Elements of the Cellular Communication System

2.1 THE GENERIC SYSTEM

The generic cellular communication system, shown in Figure 2.1, is an integrated network comprising a land-based wireline telephone network and a composite wired-wireless network.

The land-based network is the traditional telephone system in which all telephone subscribers are connected to a central switching network, commonly known as public switching telephone network (PSTN). It is a computer-based digital switching system, providing the following basic functions:

1. Switching;
2. Billing;
3. 911 dialing;
4. 1-800 and 1-900 calling features;
5. Call waiting, call transfer, conference calling, and voice mail;
6. Global connectivity;
7. Interfacing with cellular networks.

Tens of thousands of simultaneous calls can be handled by means of a single PSTN.

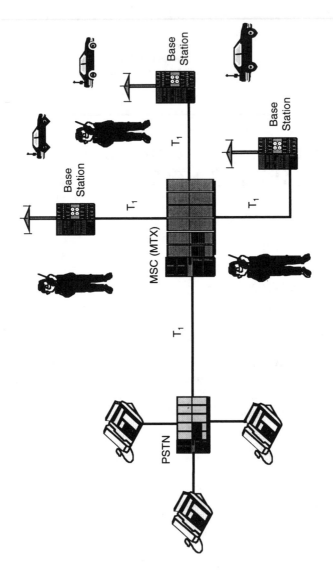

Figure 2.1 The generic telecommunication network.

The composite wired-wireless system, also shown in Figure 2.1, is the basis of today's cellular network. The heart of this system is also a digital multiplex switch, which has been derived from the PSTN by adding several functions required by the mobile phone system. It is generally identified as the mobile switching center (MSC). The function of the MSC is to:

1. Provide connectivity between PSTN and cellular base stations by means of trunks (T_1 links);
2. Facilitate communication between mobile to mobile, mobile to land, land to mobile, and MSC to PSTN networks;
3. Manage, control, and monitor various call processing activities;
4. Keep detailed records of each call for billing.

Cellular base stations are located at different convenient locations within the service area. The coverage of a base station varies from less than a kilometer to tens of kilometers, depending on the propagation environment and traffic density. For example, in a dense urban environment, typical cell radii would be less than a kilometer to a few kilometers (<5 km), whereas in a flat rural environment, the cell radii would be tens of kilometers. The system has the capacity to serve tens of thousands of subscribers in a major metropolitan area.

The mobile and portable units are generally known as *subscriber units*. A subscriber is a customer who subscribes to land-based telephone and/or cellular phone service. The cell site (base station) includes the needed cell equipment, towers, and antennas, and the supporting base station's equipment. This is the composite network that is providing modern telecommunication services.

2.2 PUBLIC SWITCHING TELEPHONE NETWORK AND ITS SUBSYSTEMS

2.2.1 System Overview

The basic telecommunication network involves terminals, a public switching telephone network, cables, and microwave links or optical fiber links as shown in Figure 2.2. A terminal can be a simple telephone set, a computer, a modem, or several workstations. The PSTN is a digitally controlled switching matrix that provides connectivity between two or more terminals. Depending on the population density, a PSTN may have well over 100,000 terminals connected to it. As a result, a major metropolitan area may have more than one PSTN switch connected to each other as depicted in Figure 2.2.

Data transmission takes place over cables, microwave links, or optical fiber links, providing global coverage via satellites, underwater cables, and optical cables.

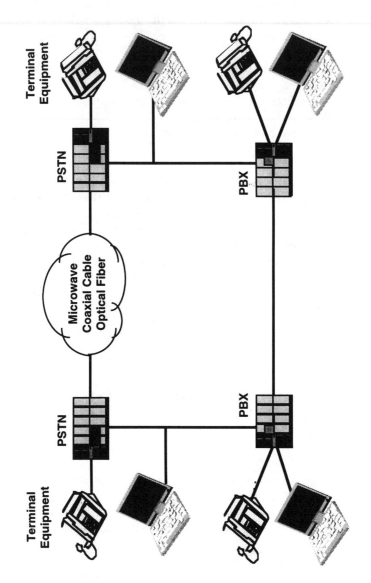

Figure 2.2 The PSTN and its subsystem, without cellular services.

Various transmission equipment and protocol converters are also used to establish global connectivity.

It is also necessary to provide telephone services to small communities such as offices and business communities in a cost-effective manner. This is accomplished by means of a small switch known as a private branch exchange (PBX), which enables an internal office call to bypass the PSTN. PBX switches are generally installed within the business subscriber's premises. Several PBXs can be installed in one or in different buildings and are connected to each other by cables.

2.2.2 Network Hierarchy

Because of widespread telecommunication services, it was necessary to group telephones around a central point (PSTN) and then group several points with higher interchange points. For long-distance dialing, each point associated with the call is classified and designated according to the highest rank of switching functions. It has five different classifications as shown in Figure 2.3. A brief description of these classes is presented in the following paragraphs.

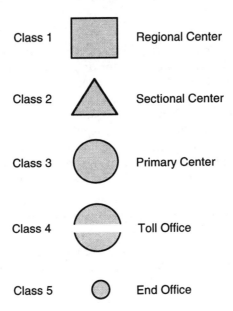

Figure 2.3 Network hierarchy.

End Office

The end office is the central switching office or the PSTN where telephone loops are terminated for switching to another telephone or a terminal. A PSTN may have well over 100,000 telephones connected to it. This is also known as a Class 5 office.

Toll Office

The switching center that handles the first stage of long-distance traffic is called the toll office. Operator-assisted calls are also handled in this location if the caller intends to do so.

Primary, Sectional, and Regional Centers

Long-distance switching, routing, and internal trunks are controlled in these locations. As an illustration let us consider a service area having four switches as shown in Figure 2.4(a). The total number of direct paths needed to interconnect these switches is given by

$$N = S(S - 1)/2$$
$$= 6 \tag{2.1}$$

where N = total number of direct paths and S = number of switches = 4. If we increase the number of switches by a factor of 2 to eight switches), the number of direct paths will be $8(8 - 1)/2 = 28$, which is not an economical number for connecting each switching office directly to each other.

A possible solution to this problem would be to use a tandem switch (transit switch) for call routing. In this solution, a regular switch is upgraded to act as a tandem switch (Figure 2.4(b)). Thus, a combination of direct and tandem (transit) links alleviates the problem of trunking. Note that a direct route is normally provided where the traffic is high and a tandem link is installed in low traffic areas.

2.3 THE MOBILE SWITCHING CENTER AND ITS SUBSYSTEMS

The MSC (Figure 2.5) is a digital cellular switching product, supporting 800-MHz cellular communication systems. It has different acronyms such as digital multiplex switch-mobile telephone exchange (DMS-MTX), mobile telephone switching office (MTSO), etc., depending on the manufacturer.

The MSC is a part of the PSTN family, designed to provide the following cellular functions:

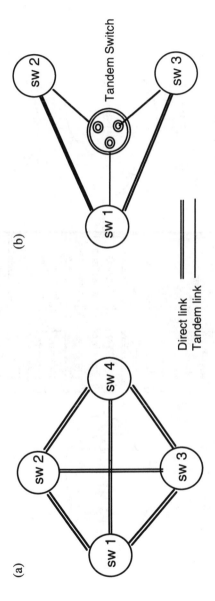

Figure 2.4 Direct trunk routing scheme: (a) service area with four switches, and (b) tandem switch.

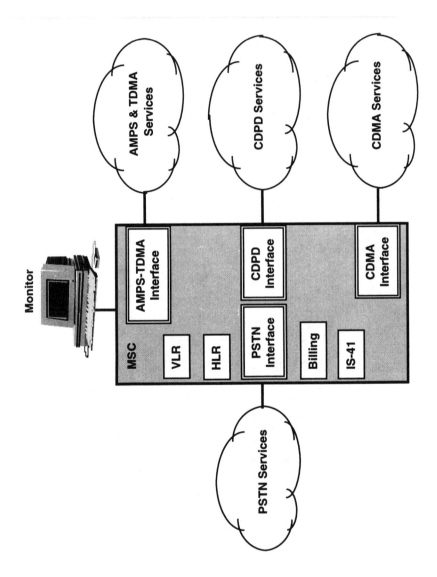

Figure 2.5 The mobile switching center.

1. Manage and control cell site equipment and connections.
2. Support multiple-access technologies such as AMPS, TDMA, CDPD, and CDMA.
3. Provide the PSTN interface.
4. Provide a visitor location register (VLR).
5. Provide a home location register (HLR).
6. Support intersystem connectivity (IS-41) [1].
7. Support call processing functions.
8. Provide billing and operation and measurement (O&M).

The MSC also provides trunking facilities (T_1 link) between the cell site and the MSC and between the MSC and the PSTN. Each of these blocks is further divided into several unique functional elements, supporting the end-to-end communication protocol specified by several FCC and CCITT standards.

2.4 CELLULAR SERVICES OFFERED BY MSC (MTX)

2.4.1 Advanced Mobile Phone Service (AMPS)

In the North American AMPS standard, EIA-IS-3-D [2], there are two separate frequency bands, adjacent to each other, designated as Band A and Band B. Each band provides 416 channel frequency pairs, having 30-kHz channel separation. Each frequency pair is isolated by 45 MHz, which permits duplex operation.

Out of 416 channels, 21 channels are designated *control channels*. Control channels are used for call setup. Twenty-one more control channels are also available if necessary. The remaining channels are used as *voice channels*. In a cellular base station, a single transceiver is used per channel (see Chapter 7 for details).

Each voice channel pair supports a single conversation at a time. Four different types of signals are transmitted over the voice channel during the course of a cellular call:

1. Voice signals;
2. Supervisory audio tone;
3. Signaling tone;
4. Data.

On the transmitting side, the voice signals are first compressed by a 2:1 syllabic compressor and then modulated by an FM modulator. On the receiving side, the incoming signal is demodulated and decompressed as 1:2 to recover the voice. This process improves the performance of FM transmission.

The supervisory audio tone (SAT) is transmitted over the forward voice channel (base to mobile) and retransmitted by the mobile back to the base. Its purpose is to indicate the continuity of the conversation. Loss of SAT indicates call completion and hand-off. Only three SAT frequencies are available (5970, 6000, and 6030 Hz). Each SAT is assigned to a cluster of cells in such a way that the same SAT is not used in a reused channel. SAT frequencies distinguish cochannel sites.

A 10-kHz signaling tone (ST) is transmitted by the mobile over the reverse voice channel to acknowledge certain commands received from the base station. This is similar to the supervisory signaling tone used in conventional telephone networks.

During hand-off, the voice channel momentarily becomes a digital channel and behaves like a control channel. A 10-Kbps data transmission takes place between the base and the mobile for channel assignment and signaling. SAT and ST tones are muted during data transmission. Once the channel assignment is complete, regular conversation resumes. About 200 ms worth of voice is muted during this process, which is heard as a "clicking" noise during a conversation.

The control channel is composed of a forward control channel (FOCC) and a reverse control channel over which data transmission takes place between the base station and mobile at 10 Kbps. Data transmission over the FOCC is based on encoding 28 control bits into a (40, 28, 5) BCH code. This code has a word length of 40 bits, 28 control bits, (40 − 28) = 12 parity bits, and Hamming distance of 5. The number of errors that can be detected by means of this encoded word is $(5 − 1)/2 = 2$. The encoded word is then transmitted by means of FSK modulation with ±8-kHz discrete level frequency deviations to represent 1 and 0. On the receiving side, the RF signal is demodulated and decoded; finally the original data are recovered.

RECC is the reverse control channel over which control data are transmitted from the mobile unit to the base station. This channel is also known as an access channel and is used by a mobile to access a land telephone or a mobile telephone. Data transmission over the RECC is based on encoding 36 control bits into a (48, 36, 5) BCH code. This code has a word length of 48 bits, 36 control bits, (48 − 36) = 12 parity bits, and a Hamming distance of 5. The number of errors that can be detected by means of this encoded word is $(5 − 1)/2 = 2$. This means that if the decoded word encounters more than two errors due to noise, interference, and fading, an alarm will be generated and the word will be declared invalid, thus reducing the RECC capacity. Both FOCC and RECC channels are full duplex and operate in a coordinated manner.

Each cell also has locate channel transceivers, as well as sectors that scan neighboring mobiles and measure the signal strengths. These signal strengths are reported to the MSC for hand-off, where hand-off is a process of changing the frequency. Its primary purpose is to assign a new frequency while a mobile moves into a new cell. This is accomplished by setting a hand-off threshold. That is, if the received signal level is too low and reaches a predefined threshold, the system controller, namely, the MSC, provides a stronger free channel (frequency) from an adjacent cell. Hand-off provides mobility, which means that a caller can move without interruption or call drop.

2.4.2 Time-Division Multiple Access (TDMA)

The North American TDMA, IS-54 [1], operates on the same frequency bands and channels specified for AMPS. The only difference is the radio itself where each voice channel is sampled and digitized to obtain a 64-Kbps PCM data. The PCM data are then compressed down to 16.2-Kbps data by means of a *vocoder*, which is then interleaved and encoded by means of a one-half rate convolutional encoder. Finally, the encoded data is modulated by means of a $\pi/4$-DQPSK modulation scheme. Data transmission is accomplished by means of a 40-ms frame structure having six time slots of 6.66 ms each (Figure 2.6(a)). Each frame supports six time slots where each mobile is assigned two time slots according to the scheme shown in Table 2.1.

Each mobile is sequentially sampled by the base station radio as shown in Figure 2.6(b). The aggregate bit rate that is processed by the base station is

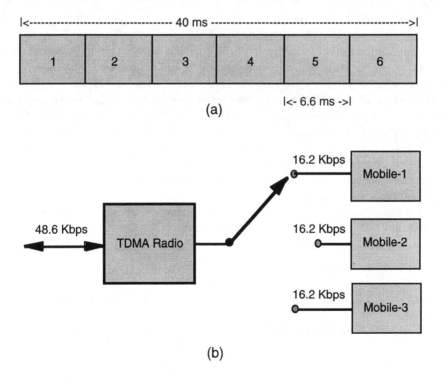

Figure 2.6 (a) TDMA frame structure. (b) Sequential sampling.

Table 2.1
TDMA Scheme

Mobile	Time Slots
1	1,4
2	2,5
3	3,6

$16.2 \times 3 = 48.6$ Kbps. All mobiles maintain synchronization by extracting timing from the base station while the base station recovers timing from the MSC.

On the reverse path, the incoming signal is demodulated, decoded by the Viterbi decoder and deinterleaved. Finally, the data is recovered by means of the vocoder.

The control channel is currently based on the existing control channels specified for AMPS. Because of capacity limitations, a new standard, IS-136 [3], is being implemented. It has the same modulation and frame structure as the voice channel.

2.4.3 Communication Protocol

The communication protocol within the cellular environment is comprised of (1) land to mobile calls, (2) mobile to land calls, and (3) mobile to mobile calls. The actual communication is based on (1) the control channel, (2) the voice channel, and (3) the locate channel. Control channels are used for paging, channel assignment, and system acquisition, and voice channels are used for voice communications. There are 21 control channels and 395 voice channels available in the North American AMPS [2]. These channels are evenly distributed among a group of cells called a *cell cluster*. Several cell clusters are available in the existing cellular communication system. The overall system capacity and performance depend largely on these factors (see Chapter 7 for details).

Various call processing features associated with this communication system are presented briefly in the following subsections.

2.4.3.1 Land to Mobile Call

Land to mobile calls are illustrated in Figure 2.7 and have these features:

- The base station transmits the channel number to the mobile over the forward control channel and turns on the assigned voice channel and a SAT tone.
- The mobile unit receives the channel number, computes the corresponding frequency, tunes to the frequency and transmits back the SAT tone as an acknowledgment.

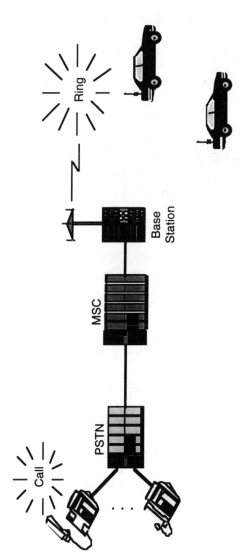

Figure 2.7 Land to mobile call.

- A 10-kHz ST ($\approx$ 0.8-sec interval) is sent by the mobile over the reverse voice channel to indicate that the call is in progress.

2.4.3.2 *Mobile to Land Call*

A mobile in idle mode continuously monitors 21 control channels and selects the strongest, usually from the nearest base station. All idle mobiles do the same after the power is on. The communication protocol is as follows (Figure 2.8):

- The land telephone subscriber dials the mobile number.
- PSTN forwards the number to the MSC.
- The MSC verifies the number and forwards the message to all the base stations connected to the MSC.
- All base stations transmit paging signals over the forward control channel.
- All mobiles in the service area receive the paging signal.
- The wanted mobile responds to the page by sending an identification number over the reverse control channel.
- The base station informs the MSC that the mobile is available.
- The MSC assigns a free voice channel and informs the base station.

Mobile to land communication protocol supports call origination from a mobile, described as follows:

- Mobile power "ON," scans 21 control channels, locks on to the strongest control channel.
- The mobile subscriber enters the desired number and presses the "SEND" key.

At this point, the following information is transmitted to the base station over the reverse control channel:

- MIN1: Mobile identification number; a seven-digit telephone number, translated into 24 bits.
- MIN2: Mobile identification number; a three-digit area code, translated into 10 bits (if the mobile is outside home territory).
- ESN: Electronic serial number; every mobile has an ESN, which is programmed during factory assembly.
- The base transmits the information to the MSC.
- The MSC coordinates the connectivity by assigning a free channel and a SAT to the mobile, and by sending the telephone number to the PSTN.
- The PSTN provides the land connectivity.
- The base station provides the channel and the SAT tone to the mobile.

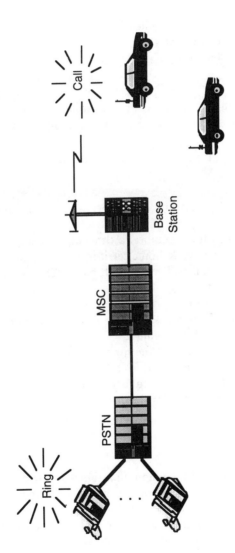

Figure 2.8 Mobile to land call.

- The mobile tunes to the frequency, loops back the SAT over the voice channel, and communication begins.

2.4.3.3 Mobile to Mobile Call

Note that in mobile to mobile communication, there is no direct link between the mobiles; the radio link is always between the cell site and the mobile (Figure 2.9). The call is processed by the MSC; PSTN is *not* involved in this communication protocol.

- All mobiles continuously scan 21 control channels.
- A mobile subscriber enters the desired number and presses the "SEND" key.

The following information is transmitted to the base station over the reverse control channel:

- MIN1: Mobile identification number; a seven-digit telephone number, translated into 24 bits.
- MIN2: Mobile identification number; a three-digit area code, translated into 10 bits (if the mobile is outside home territory).
- ESN: Electronic serial number. Every mobile has an ESN, which is programmed during factory assembly.
- The base transmits the information to the MSC.
- The MSC finds two idle channels, one for mobile 1 and the other for mobile 2, and informs the base of its findings.
- The base station in turn informs the mobiles over forward control channels, and then turns on the channels and SAT.
- The mobiles tune to the assigned frequencies, loop back the SAT tone over the voice channel, and communication begins.

2.4.3.4 Call Supervision

Each active channel is sampled by a scanning a signal strength measuring receiver, known as a locate receiver, which is located at the base station. A set of signal strengths is then transmitted to the MSC for further processing such as readings on the relative signal strengths, and presence of the SAT. The signal strengths are used to direct a mobile to another cell if the receiving signal strength is too weak with respect to a predetermined threshold. This mechanism is known as *hand-off*. Absence of a SAT is an indication of hand-off or a call drop. Each cell site has at least one locate receiver if it is an OMNI site, three locate receivers for a three-sector site,

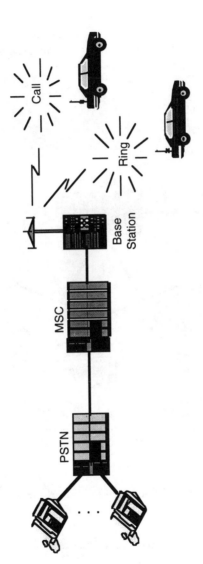

Figure 2.9 Mobile to mobile call.

and six locate receivers for six sector sites. Call supervision lasts for the full duration of a call.

2.4.3.5 Hand-Off

Hand-off is a process of transferring a call from one cell to another without service interruption (Figure 2.10). The basic concept of the process is presented in the following list:

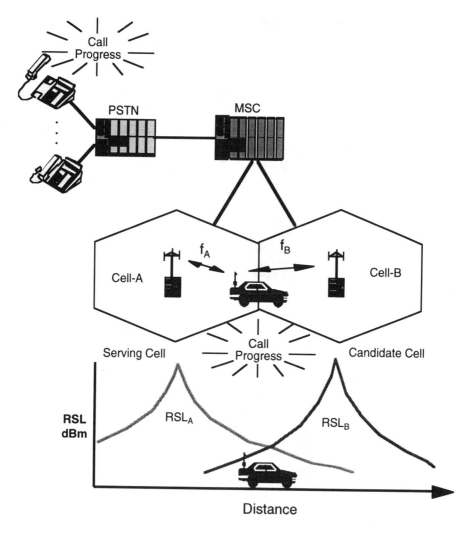

Figure 2.10 The basic hand-off process.

- A mobile in the serving cell (Cell A) is in conversation with a land telephone over channel frequency f_A.
- As the mobile moves towards Cell B (the candidate cell), its received signal level (RSL) from the serving cell drops. The RSL is monitored by a locate receiver in the cell. Every cell has a locate receiver for this purpose.
- When the RSL drops below a threshold, determined by the service provider, the base station informs the MSC of this occurrence. The MSC asks all the adjacent base stations to measure the signal strength of the mobile in question.
- The MSC receives all RSL values from adjacent bases and keeps track of the statistics.
- When the mobile's signal becomes stronger in an adjacent base, the MSC arranges a new channel (f_B) from the candidate cell and informs the serving base of this action. The serving base sends a digital message over the current voice channel advising the mobile to switch to frequency f_B, which belongs to the candidate cell, Cell B.
- The MSC also arranges for this new channel to switch to the land telephone.
- The entire process requires about 200 ms. The voice is muted during this period, which may be heard as a "click" by human ears.

2.4.4 Cellular Digital Packet Data (CDPD)

Although an existing cellular channel can be used to transmit fax or computer data with the aid of a cellular modem, it is not suitable for services generally offered by the Internet, because of the short and bursty nature of the transmission of packet data through the Internet. To alleviate this problem, the U.S. cellular industries have come up with a standard for packet data known as cellular digital packet data (CDPD) [4]. The existing cellular infrastructure is used for this service where a free AMPS channel can be used as a CDPD channel.

Data transmission over the CDPD channel is based on encoding 47 data bits into a (63,47) Reed-Solomon code which can correct as high as $(63 - 47)/2 = 8$ errors. The encoded word (block) is then transmitted by means of GMSK modulation at 19.2 Kbps. Note that in GMSK modulation, the power spectral density is such that more than 90% energy is retained within a transmission bandwidth of $19.2 \times 1.2 \approx 23$ kHz, which is sufficient for the available 30-kHz channel spacing. On the receiving side the RF signal is demodulated, decoded, and the original data finally recovered. Because of error control coding, the CDPD channel offers highly reliable data transmission over the cellular channel.

2.4.5 Code-Division Multiple Access (CDMA)

North American CDMA operation, IS-95 [5], uses the existing 12.5-MHz cellular band. Ten different frequency bands of 1.25 MHz each are derived for CDMA operation. Each of these bands can support 64 pseudonoise (PN) codes (Walsh code). CDMA air-link is based on a forward link and a reverse link, separated by 45 MHz. The forward link has four different channels:

1. *Pilot channel:* Used by the mobile to acquire phase, timing and signal strength.
2. *Paging channel:* System parameters are transmitted by the base station.
3. *Sync channel:* Mobile synchronizes by means of this channel.
4. *Traffic channel:* Mobile and base communicate over this channel.

The reverse link has two communication channels:

1. *Access channel:* Used by the mobile for system access, traffic channel request, call origination, page response, and system registration.
2. *Traffic channel:* Used by the mobile for voice communication, command response, and information request from the base.

CDMA transmission is based on the spread-spectrum (SS) modulation technique in which many users have access to the same band. This is accomplished by assigning a unique orthogonal code (Walsh code) to each user. This code is then directly MOD-2 added with the corresponding base-band data as illustrated in Figure 2.11. Since the PN code is high-speed data compared to the base-band data, the power spectrum associated with the composite binary data is much wider than

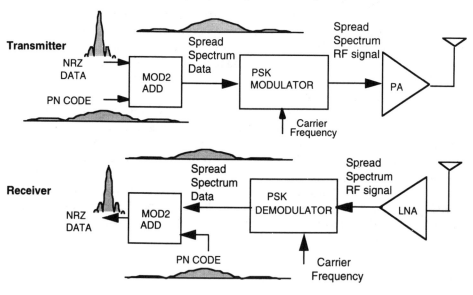

Figure 2.11 DS-CDMA spreading and despreading concept.

that of the base-band data. These high-speed composite data are then modulated, amplified, and transmitted by a carrier frequency, common to all users.

On the receiving side the weak signal is amplified, demodulated, and despread to recover the original signal. This process only works if the transmitting-receiving PN codes are identical. See Chapter 4 for details.

2.4.6 Other Services Offered by MSC (MTX)

2.4.6.1 Home Location Register (HLR)

The HLR is a fixed database for storage and management of subscriber information. It provides and stores the following information:

- Subscriber data such as subscriber service profiles;
- Information on the subscriber locations;
- Information on subscriber service status;
- Mobile identification number;
- Directory number;
- Electronic serial number;
- Current registration of a mobile;
- Call duration of a mobile;
- Information pertaining to service denial;
- Supplementary data on call origination/termination;
- Information on long-distance routes a mobile has chosen;
- Information about the service provider.

The HLR communicates with the visitor location register (VLR) and supports IS-41 messages from the VLR indicating that a subscriber has registered in the system.

2.4.6.2 Visitor Location Register (VLR)

The VLR is a dynamic database used to store roaming mobile subscriber information associated with the MSC. The VLR contains all subscriber data required for the call handling of mobile subscribers currently located in the area. the VLR stores temporary subscriber information, which can change as a result of normal operation of the system. the VLR communicates with the HLR through an internal connection.

2.4.6.3 IS-41 Networking

International Standard 41 (IS-41) [6] is a special link protocol that enables different mobile switching centers to communicate with each other. The IS-41 link protocol enables roaming, that is, a roamer can originate a call from a different service area, supported by a different service provider's MSC. Mutual agreement is necessary to provide this service.

2.4.6.4 Billing

The MSC also keeps detailed records of each call, including the following:

- Air-link usage;
- Trunk (T_1) usage;
- Release information.

This information is recorded and stored for billing purposes. Detailed statistics on errors, delays, and volume of traffic are also recorded and can be retrieved for system performance analyses.

2.4.6.5 Mobile Tracking

The MSC is also capable of tracking an active mobile on AMPS/TDMA or a CDMA call, and it can trace all the call processing activities. This feature is useful for mapping the AMPS/TDMA/CDMA cell's RF coverage for cell site optimization.

2.5 THE MOBILE

The mobile or subscriber unit (Figure 2.12) is a portable voice and/or data communication transceiver, designed to communicate with cell site radios in any of the allocated channels. It operates in the full-duplex mode having a forward path and a reverse path. There is 45-MHz isolation between the forward and the reverse path, which protects the channel from mutual interference.

Each mobile has a 10-digit telephone number, represented by a 34-bit mobile identification number (MIN), which is used for both transmitting and receiving calls. The MIN is divided into two parts, MIN1 and MIN2. MIN1 is represented by a 7-digit telephone number, translated into 24 bits. MIN2 is represented by a 3-digit area code, translated into 10 bits. For example, the telephone number (214) 684-5981 would be represented as shown in Table 2.2.

Each mobile is also permanently programmed at the factory with a 32-bit electronic serial number (ESN), which guards against tampering.

2.6 THE CELL

The cell is a geographical area covered by RF signals. The RF source is located at the center of the cell as shown in Figure 2.13. This is essentially a radio communica-

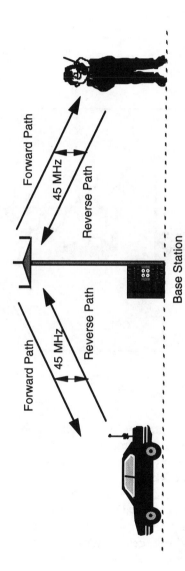

Figure 2.12 A mobile cellular communication scenario.

Table 2.2
MIN Representation
of a Common Telephone
Number

MIN2	MIN1
214	684-5981
10 bits	24 bits

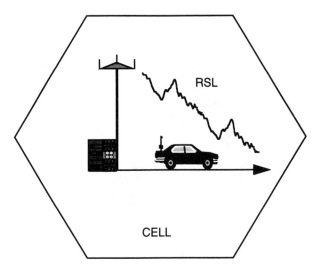

Figure 2.13 The cell.

tion center comprising radios, antennas, and much supporting equipment enabling communication between mobile to land, land to mobile, and mobile to mobile units. The entire communication process is controlled and monitored by the system intelligence resident within the MSC.

The shape and size of the cell depends on several parameters such as ERP, antenna radiation pattern, and propagation environments. Traditionally, a practical cell is assumed to be highly irregular having regular RSL at the cell boundary. On the other hand, the analytical cell, generally used for planning and engineering, is assumed to be a perfect hexagon as shown in Figure 2.13. Consequently, a discrepancy arises between the analytical cell and the practical cell. As mentioned the analytical cell is used for system planning and design, and its initial deployment is based on computer-aided prediction tools (see Chapter 5) that closely approximate a practical cell in a given propagation environment. Traffic engineering (see Chapter 6) also plays an important role in determining the size of the cell.

2.7 CELL COVERAGE

Cell coverage primarily depends on user-defined parameters such as transmitting power, antenna height, antenna gain, antenna location, and antenna directivity. Several other parameters such as propagation environment, hills, tunnels, foliage, and buildings greatly affect the overall RF coverage. These types of parameters are not user defined, vary from place to place, and are difficult to predict. As a result, a practical cell is highly irregular in multipath environment as depicted in Figure 2.14.

Consequently, several prediction models have been developed in recent years. The two most widely used propagation models, accommodating most of these anomalies of propagation, are the Okumura-Hata and Walfisch-Ikegami propagation models. The foundation of most computer-aided prediction tools available today is also based on these models. These prediction models are based on extensive experimental data and statistical analyses that enable us to compute the received signal level in a given propagation medium.

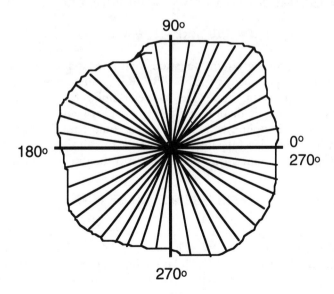

Figure 2.14 A practical cell having different coverage in different directions due to multipath, shadowing, hills, vegetation, foliage, and building clutter factors.

A detailed analysis of these models is presented in Chapter 5, where we conclude that the path loss characteristics follow the familiar equation of a straight line of the following form:

$$L \text{ (dB)} = L_o \text{ (dB)} + 10\gamma \log(d/d_o) \tag{2.2}$$

where

d_o = Fresnel zone break point $(d_o \approx 4h_1h_2/\lambda)$

d = coverage in a particular direction

h_1 = base station antenna height

h_2 = mobile antenna height

where the coverage is d_0

γ = propagation constant in the same direction (function of environment)

L_o = intercept (function of environment, antenna height, location, etc.), dB

L = path loss in the same direction.

With RSL being the received signal level, we predicted that

$$d \approx d_o 10 \text{ (ERP} - L_o - \text{RSL)}/10\gamma \tag{2.3}$$

which indicates that for a given propagation environment and cell site location, the coverage depends on parameters such as ERP, RSL, and antenna height, which are user defined, and on several clutter factors, determining L_o, which is the intercept. Consequently, coverage prediction and cell site deployment, classified as RF engineering, is a major discipline within the cellular industries. It is also an ongoing process even in a fully developed cellular system for cell site optimization, performance, and capacity enhancement. See Chapters 5, 6, and 7 for details.

2.8 CELL CLUSTER AND FREQUENCY REUSE PLAN

A cell cluster is a group of identical cells in which all of the available channels (frequencies) are evenly distributed. The most widely used plan is the $N = 7$ cell cluster (Figure 2.15(a)) where 416 cellular channels are evenly distributed among 7 cells having approximately 59 channels per cell, which then repeats itself over and over according to Figure 2.15(b). Because of the limited number of channels, their reuse must be carefully planned. In hexagonal geometry, this reuse plan is given by [7–9]

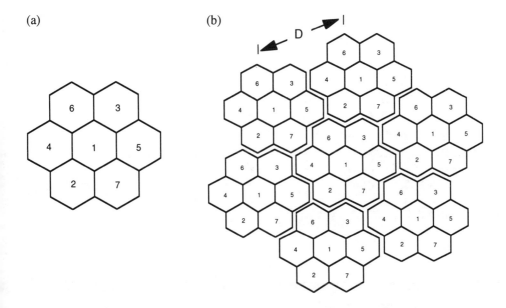

Figure 2.15 (a) A seven-cell cluster. (b) A seven-cell reuse plan.

$$\frac{D}{R} = \sqrt{3N} \tag{2.4}$$

where D is the reuse distance, R is the cell radii, and N is the modulus.

A number of other well-established plans are available for cellular applications. These patterns are based on $N = 3, 4, 7, 9,$ and 12 (see Chapters 7 and 8 for details). Column 2 of Table 2.3 provides a brief summary of D/R ratios as a function of the modulus N.

2.9 COCHANNEL INTERFERENCE

A cochannel interferer has the same nominal frequency as the desired frequency. It occurs as a result of multiple uses of the same frequency. A cell site, radiating in all directions (OMNI site), is represented by a carrier-to-interference ratio as follows:

$$C/I = 10 \ \log[1/j\,(D/R)^{\gamma}] \tag{2.5}$$

Table 2.3
Capacity Performance As a Function of the D/R Ratio for an OMNI Site

N	$\dfrac{D}{R} = \sqrt{3/N}$	$\dfrac{C}{I} = 10 \log\left[\left(\dfrac{1}{j}\right)\left(\dfrac{D}{R}\right)^{\gamma}\right]$	*Channel Capacity per Cell* $\left(\dfrac{416}{N}\right)$
3	3	−11	138
4	3.46	−13	104
7	4.58	−18	59
9	5.19	−20	46
12	6	−23	34

where

j = number of cochineal interferers (j = 1, 2, . . ., 6)

γ = propagation constant

D = frequency reuse distance

R = cell radii.

With γ = 4, repeat distance = 4.58, and j = 6, the cochannel interference becomes,

$$C/I = 18.6 \text{ dB} \quad \text{Six interferers (worst case)} \quad (2.6)$$

Column 3 of Table 2.3 summarizes the results for other frequency plans.

2.10 CHANNEL CAPACITY

Channel capacity is determined by *traffic engineering* (see Chapter 6), which deals with the provision of communication circuits in a given service area, for a given number of subscribers, with a given grade of service (GOS).

GOS is essentially the call blocking probability, defined in terms of the number of calls that will be blocked during the busy hour due to lack of channels. A GOS of 2% is generally used for North American cellular services.

Another design parameter is the average call holding time (ACHT). ACHT is the average duration of a call during which each caller is expected to hold during the busy hour. The call holding time varies depending on the type of subscriber; that is, business, private, etc. Typical call holding times vary between 120 and 180 sec.

Cell capacity is then determined by using the traffic table (Erlang B table) for a given grade of service. This table provides traffic engineers with the appropriate number of trunks (channels) required to provide a particular grade of service for a given number of subscribers. Column 4 in Table 2.3 provides a set of cell capacities for a selected set of frequency plans.

2.11 SECTORIZATION

The 120-deg sectorization is achieved by dividing a cell into three sectors of 120 deg each, as shown in Figure 2.16(a). Directional antennas are used in each sector for a total of three antennas per cell. The 60-deg sectorization is achieved by dividing a cell into six sectors of 60 degrees each, as shown in Figure 2.16(b). Each sector is treated as a logical OMNI cell; directional antennas are used in each sector for a total of six antennas per cell. Because of the narrow antenna beam width, channels can be repeated more often, thus enhancing the capacity. This configuration is generally used in dense urban environments.

Because of directionalization, the *C/I* equation presented in (2.5) is now modified as

$$\frac{C}{I} = 10 \log \left[\left(\frac{1}{j} \right) \left(\frac{\phi_1}{\phi_2} \right) \left(\frac{D}{R} \right)^\gamma \right] \tag{2.7}$$

where ϕ_1/ϕ_2 is the antenna directivity factor. Other parameters in (2.7) are defined in (2.5). With a typical antenna directivity factor of 10, we see that *C/I* for $N = 3$ and $N = 4$ of Table 2.3 now achieves our performance objectives. The result is summarized in Table 2.4 for convenience.

(a)

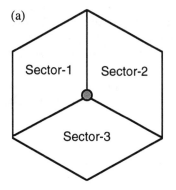

(b)

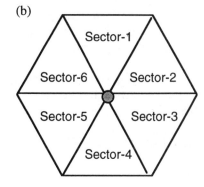

Figure 2.16 Sectorization: (a) three-sector configuration, and (b) six-sector configuration.

Table 2.4
Capacity Performance as a Function of the D/R Ratio for a Sectored Site

N	$\dfrac{D}{R} = \sqrt{3/N}$	$\dfrac{C}{I} = 10 \log\left[\left(\dfrac{1}{j}\right)\left(\dfrac{\phi_1}{\phi_2}\right)\left(\dfrac{D}{R}\right)^{\gamma}\right]$ (dB)	Channel Capacity per Cell $\left(\dfrac{416}{N}\right)$
3	3	−21	138
4	3.46	−23	104
7	4.58	−28	59
9	5.19	−30	46
12	6	−33	34

2.12 CELL SITE CONFIGURATIONS

A cell site (Figure 2.17) supports several radios, one radio per frequency (channel). The typical number of radios per cell site is approximately 48, according to North American cellular standard AMPS. Each AMPS radio provides services to one subscriber at a time. On the other hand, a TDMA radio can support three subscribers in a TDMA manner, thus increasing channel capacity threefold.

To operation, a cell site uses a base-band bipolar signal, received from the T_1 link. This signal is converted into NRZ data, demultiplexed, and fed into the radio transmitter for modulation, upconversion, and amplification. These radio signals are then combined to form a single stream of high-power radio signals, filtered by means of a transmit bandpass filter (duplexer), and transmitted through the main antenna.

On the receiving path, the incoming radio signal is filtered (duplexer) and fed to the corresponding radio receivers through the splitter. The function of the splitter is to amplify the incoming signal by means of a low-noise amplifier (LNA), and filter and split it into an appropriate number of receive signals, determined by the number of radios on the cell site. An identical receiver path, providing space diversity, recovers the same signal. Both receive signals are then compared for the strongest signal.

Finally, the composite multiple signals are placed in the appropriate time slots by means of the multiplexer, converted into a bipolar signal, and transmitted to the MSC through T_1 links.

2.13 CELLULAR ANTENNAS

An antenna is a signal processing device that transmits and receives electromagnetic signals at the same time. It is available in two general categories: (1) passive antennas and (2) active antennas.

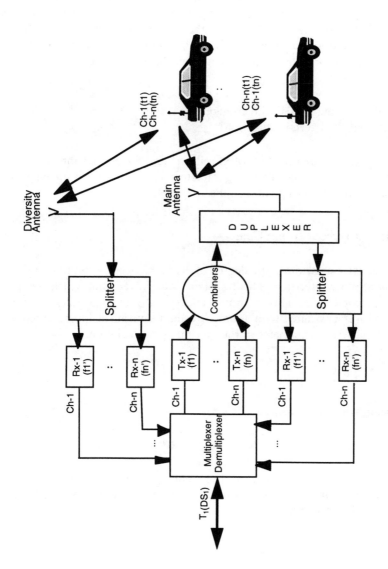

Figure 2.17 Cell site configuration.

The radiation pattern of a passive antenna depends on the type and construction of the device because the radiation pattern is not fixed until after construction of the device. However, it can be guided to a certain degree by mechanical means. Mechanical downtilt is a common practice to control the signal within a cell.

The radiation pattern of an active antenna depends on the type, construction, and built-in signal processing technique of the device. Generally, digital signal processing techniques are used to generate a desired radiation pattern. The radiation pattern can be steered in a given direction as well. Also, there are two general classes of radiation pattern:

1. OMNI directional (in all directions);
2. Directional (in a certain direction).

OMNI antennas are used in OMNI cell sites and directional antennas are used in sectored sites. Some of the antenna parameters, essential for cell site engineering, are described in the following subsections.

2.13.1 Antenna Directivity and Gain

Antenna directivity determines the degree of energy concentration in one direction with respect to other directions. This translates into power gain, which is expressed relative to isotropic gain. This is shown conceptually in Figure 2.18, where the antenna pattern is associated with a main lobe and several sidelobes. Antenna gain is an important design parameter in the link budget calculation.

2.13.2 Antenna Beam Width

Antenna beam width is measured as:

$$\text{Beam width} = 2\theta \tag{2.8}$$

where θ is the angle with respect to bore sight where the voltage is 0.707 of its maximum value (Figure 2.19). The performance of a sectored cell site largely depends on the antenna beam width.

2.13.3 Antenna Front-to-Back Ratio

The front-to-back ratio is defined as the ratio of the power radiated from the main lobe to that of the back lobe (Figure 2.20). This is defined as:

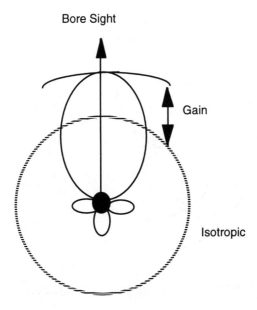

Figure 2.18 Antenna directivity and gain.

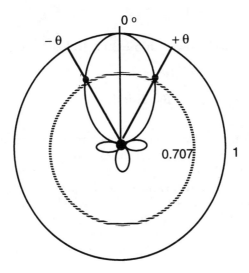

Figure 2.19 Antenna beam width.

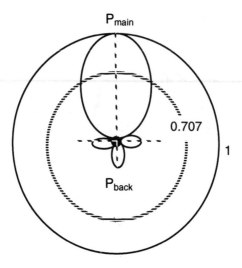

Figure 2.20 Front-to-back ratio.

$$\text{Front to back ratio} = 10 \, \log\frac{P_{\text{main lobe}}}{P_{\text{back lobe}}} \tag{2.9}$$

For example,

$$\text{Front to back power ratio} = 30 = 10 \, \log(30) = 14.8 \text{ dB}$$

2.13.4 Frequency Response and Bandwidth

Every antenna has a frequency response, which means that it passes certain bands of frequencies and attenuates other frequencies. This is conceptually shown in Figure 2.21 where bandwidth (BW) = $(f_H - f_L)$, where f_H is the upper 3-dB frequency, f_L is the lower 3-dB frequency, and f_o is the center frequency.

The selectivity is given by $Q = (f_H - f_L)/f_o$, which is determined by the voltage standing wave ratio (VSWR). Typical antenna parameters for 800-MHz cellular sites are listed in Table 2.5.

The antenna is an essential component in all radio communication systems. For cellular applications, antenna type, frequency response, radiation pattern, tower height, antenna downtilt, etc., are essential engineering problems that require proper attention to RF link design. (See Chapter 8 for details.)

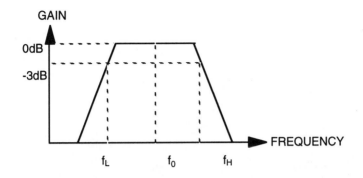

Figure 2.21 Frequency response.

Table 2.5
Typical Antenna Parameters

Antenna Type	Frequency Response (MHz)	Bandwidth (MHz)
Main antenna (Tx/Rx)	824–894	70
Diversity antenna (Rx)	824–849	25
Mobile antenna (Tx/Rx)	824–894	75

References

[1] IS-54, "Dual-Mode Mobile Station–Base Station Compatibility Standard," PN-2215, Electronic Industries Association Engineering Department, December 1989.

[2] Interim Standard, "Cellular System Mobile Station–Land Station Compatibility Specifications," IS-3-D, Electronic Industries Association Engineering Department, March 1987.

[3] IS-136, "TDMA Cellular/PCS-Radio Interface-Mobile Station–Base Station Compatibility–Digital Control Channel," Revision A, PN-3474.1, Electronic Industries Association Engineering Department, November 17, 1995.

[4] A group of U.S cellular industries, "CDPD," Release 1.0, July 1993.

[5] IS-95, "Mobile Station–Base Station Compatibility Standard for Dual-Mode Wideband Spread Spectrum Cellular Systems," TR-45, PN-3115, Electronic Industries Association Engineering Department, March 15, 1993.

[6] IS-41, Interim Standard, "Cellular Radio Telecommunications Intersystems Operation, Functional Overview," NP EIA/TIA IS-41.1-B, Electronic Industries Association Engineering Department, December 1991.

[7] Mehrotra, Asha, *Cellular Radio, Analog and Digital Systems,* Norwood, MA: Artech House, 1994.

[8] Mehrotra, Asha, *Cellular Radio Performance Engineering,* Norwood, MA: Artech House, 1994.

[9] Lee, William C. Y., *Mobile Cellular Telecommunications Systems,* New York: McGraw-Hill Book Company, 1989.

CHAPTER 3

▼▼▼

NORTH AMERICAN
DUAL-MODE AMPS-TDMA

3.1 INTRODUCTION

The North American dual-mode AMPS-TDMA standard [1,2] is a narrowband mobile cellular system, where one mode of operation is AMPS, and the second mode of operation is TDMA. The dual-mode cell site configuration, shown in Figure 3.1, is based on a programmable dual-mode radio and supporting equipment. The personality of the radio (analog or digital) is software controlled, offering both analog and digital services from the same cell site.

The cell site is a multiple-access network in which each channel is individually modulated by the respective radio, upconverted, amplified, and combined to form a high-power channel group. The composite RF signal is then fed to an antenna and transmitted. On the receiving path, the incoming weak RF signal is amplified by a low-noise amplifier, spilt into several signals, and fed into the respective radio receivers for demodulation and signal recovery.

An identical receiving path provides space diversity, which receives the same signal through a separate antenna. The two receive signals are then compared and the strongest signal is selected; this is a built-in feature within the radio. The cell site is connected to the mobile switching center through a cross-point switch via a T_1 link. The cross-point switch also converts T_1 data from serial to parallel and

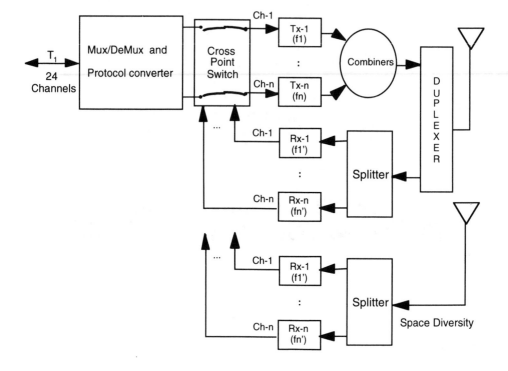

Figure 3.1 Dual-mode cell site configuration based on dual-mode radio (analog and digital).

parallel to serial format. Radio port assignments are performed during cell site engineering. Once the cell site is configured, the radio port assignment cannot be changed dynamically. The entire communication process is controlled and monitored by the system intelligence, resident in the mobile switching center (**MSC**). A brief functional description of the MSC was provided in Chapter 2.

3.2 THE AMPS

The AMPS mode of operation is based on FDMA (Figure 3.2), in which each FDMA channel is used by a single mobile via FM transceivers. This is accomplished by dividing the 12.5-MHz band into 416 narrowband FDMA channels of 30 KHz each.

Of the 416 channels, 21 channels are used as control channels. The remaining 395 channels are used as voice channels. AMPS is a full-duplex communication system, which means that simultaneous transmission takes place in both directions, identified as (1) forward path or downlink transmission and (2) reverse path or uplink transmission as shown in Figure 3.3.

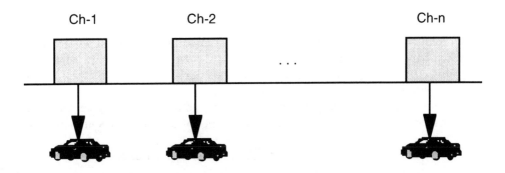

Figure 3.2 FDMA technique used in AMPS where voice communication is based on FM and call setup is based on FSK.

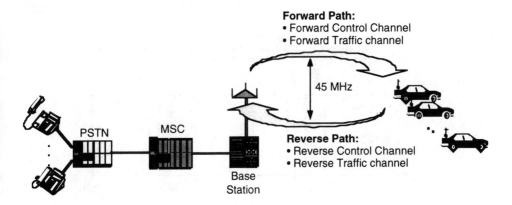

Figure 3.3 The AMPS full-duplex operational scheme.

The forward path is comprised of two communication channels: (1) the forward control channel and (2) the forward voice channel. Similarly, the reverse path consists of two channels: (1) the reverse control channel and (2) the reverse voice channel. The control channels are used for channel assignment, paging, messaging, etc., and voice channels are used for conversations. A 45-MHz guardband is provided to avoid interference between forward and reverse channels as indicated in Figure 3.3.

An additional radio known as a locate receiver is used in the base station to locate mobiles within a cell or a sector. This radio is used as a scanning receiver where the transmitter is disabled. It is used to measure received signal strengths from certain mobiles upon receiving a command to do so from the MSC. The measured signal strengths are then used to determine a candidate cell for a possible call transfer (hand-off).

3.3 AMPS FREQUENCY BANDS AND CHANNELS

AMPS frequency bands, allocated by the regulating authority, are shown in Figure 3.4. Each band (Band A and Band B) occupies 12.5 MHz, of which 10 MHz is in the nonexpanded spectrum (NES) and 2.5 MHz is in the expanded spectrum (ES) [1].

There are 333 channels per band in NES and an additional 83 channels in ES for a total of 416 channels in each band. The types of channels are summarized in Table 3.1.

Frequency assignment is related to the channel number as follows:

$$\text{Transmit frequency} = (0.03N + 870) \text{ MHz:} \qquad \text{(NES)}$$
$$= 0.03(N - 1023) + 870 \text{ MHz:} \quad \text{(ES)} \qquad (3.1)$$

$$\text{Receive frequency} \ = 0.03N + 825 \text{ MHz:} \qquad \text{(NES)}$$
$$= 0.03(N - 1023) + 825 \text{ MHz:} \quad \text{(ES)} \qquad (3.2)$$

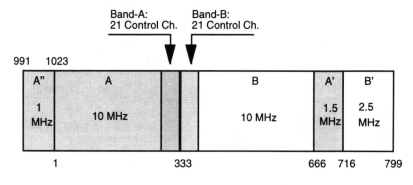

A = Band-A, Non-Expanded Spectrum   B = Band-B, Non-Expanded Spectrum
A' = Band-A Expanded Spectrum   B' = Band-B, Expanded Spectrum
A" = Band-A, Expanded Spectrum

Figure 3.4 Allocated bands for an 800-MHz cellular system.

Table 3.1
Channel Numbers

System	Type	BW (MHz)	Number of Channels
A″	ES	1	33
A	NES	10	333
B	NES	10	333
A′	ES	1.5	50
B′	ES	2.5	83

where N is the channel number ($N = 1, 2, \ldots, 1023$). Thus by knowing the channel number, the associated pair of frequencies can be obtained from the preceding equations. Detailed information on frequency planning, cell planning, and associated performance issues is presented in Chapter 7.

The communication between the base station and the mobile is based on a special call processing protocol, described in the IS-54 standard [1]. A brief description of this process is given in the following section.

3.4 AMPS CALL PROCESSING

3.4.1 Introduction

There are 21 control channels and 395 voice channels in each band. The control channels are located between Band A and Band B as shown in Figure 3.5. As mentioned, control channels are used for channel assignment, paging, and messaging, and voice channels are used for conversations. Voice channels are also used intermittently for hand-off while the call is in progress.

The basic cellular call processing involves these types of calls:

1. Land to mobile calls;
2. Mobile to land calls;
3. Mobile to mobile calls;
4. Hand-off.

These call processing functions are discussed in the following sections.

3.4.2 Land to Mobile Call

A mobile in idle mode scans 21 control channels (Figure 3.6(a)) and selects the strongest forward control channel (Figure 3.6(b)), usually from the nearest base station. All idle mobiles do the same after the power is on. The mobile then monitors the forward control channel for a possible paging message if any. If the mobile recognizes the page, it waits for the busy/idle bit to turn "idle' and then responds to the paging message to indicate its presence. The system then sends a voice channel assignment over the forward control channel; the mobile acknowledges over the reverse control channel and computes the corresponding frequency by means of (3.1) and (3.2). It then tunes to the frequency (Figure 3.6(c)) and conversation begins. The sequence of events that takes place between the MSC, base station, and mobile are summarized in Table 3.2.

3.4.3 Mobile to Land Call

User turns the mobile power on. After a power ramp-up time, the mobile goes into an idle mode, scans 21 control channels (Figure 3.6(a)) and selects the strongest

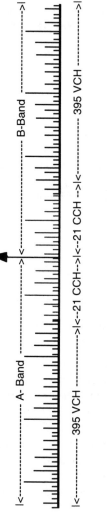

Figure 3.5 AMPS channels, illustrating the relative position of control channels and voice channels.

forward control channel (Figure 3.6(b)), usually from the nearest base station. The user dials the number and presses the SEND button. The mobile awaits for the busy/idle bit to turn "idle" and sends an access request message over the reverse control channel. The system then assigns a voice channel over the forward control channel; the mobile acknowledges over the reverse control channel, computes the corresponding frequency, tunes to the computed frequency (Figure 3.6(c)), and the conversation begins. This sequence of events is summarized in Table 3.3.

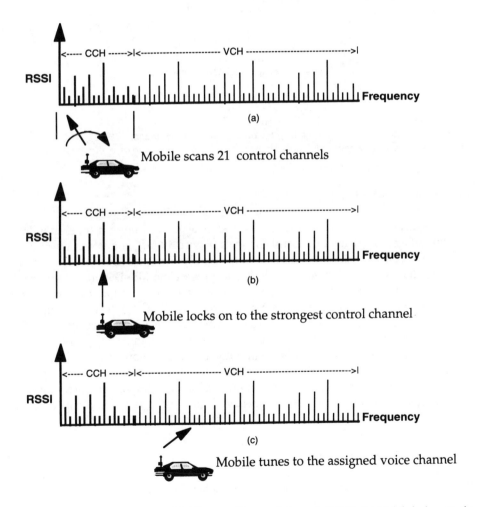

Figure 3.6 Basic call processing. (a) Mobile scans 21 control channels (CCH). (b) Mobile locks onto the strongest control channel. (c) Mobile receives a voice channel assignment command from the MSC, over the CCH, tunes to the assigned voice channel (VCH), and conversation begins.

Table 3.2
RECC Message Structure

WORD-A	An address word. It is always sent to identify the mobile.
WORD-B	An extended address word. It is sent on request from the base station to identify a roamer.
WORD-C	Serial number. Every mobile has a unique serial number provided by the manufacturer. It is used to validate each user.
WORD-D	First word of the called number.
WORD-E	Second word of the called number.

Table 3.3
Types of Messages Sent Over the RECC

Page response	Mobile receives a page from the base station and responds.
Call origination	Mobile originates a call.
Order confirmation	Mobile responds to the order from the base station.
Order message	Mobile sends an order message to the base station to carry out certain functions.

3.4.4 Mobile-to-Mobile Call

The user turns the power on to mobile 1. After a power ramp-up time, mobile 1 goes into idle mode, scans 21 control channels (Figure 3.6(a)), and selects the strongest forward control channel (Figure 3.6(b)), usually from the nearest base station. The user dials the number for mobile 2 and presses the SEND button. Mobile 1 waits for the busy/idle bit to turn "idle" and sends an access request message over the reverse control channel. The system then pages mobile 2, which is assumed to be in idle mode, locked into the strongest control channel in another cell. Mobile 2 receives the paging message over the forward control channel and responds over the reverse control channel. The system assigns a voice channel to mobile 1 and another voice channel to mobile 2, thus providing the connectivity. Both mobiles acknowledge over the respective reverse control channels, compute the corresponding frequency, tune to the computed frequency (Figure 3.6(c)), and the conversation begins.

3.4.5 Basic Hand-Off

During a call, if a subscriber is moving from a serving cell to an adjacent cell, the mobile unit's voice channel signal strength declines and becomes a hand-off candidate. The system intelligence within the MSC receives a signal strength measurement from the locate receiver over the T1 link. These signal strengths belong to adjacent target cells. The MSC then makes a hand-off decision by selecting the best cell, finding a free channel, and advising the base station of its choice. The base station removes the SAT from the voice channel, indicating that a channel assignment is imminent. At the same time, the base station momentarily changes the modulation scheme from FM to FSK and turns the analog voice channel into a digital channel, similar to a control channel. A new voice channel assignment then takes place over the forward voice channel (*now acting as a forward control channel*). The mobile acknowledges over the reverse voice channel (*now acting as a reverse control channel*), computes the corresponding frequency and tunes to it, and the conversation continues with a short mutation of 100 to 200 ms. This mechanism is illustrated in Figure 3.7.

3.5 CELLULAR CONTROL CHANNEL

3.5.1 Introduction

The cellular control channel (CCC) is composed of a forward control channel (FOCC) and a reverse control channel (RECC) over which data transmission takes place between the base station and the mobile unit (Figure 3.8). Channel separation between the FOCC and RECC is 45 MHz. Control channels, used to set up calls, are also known as setup channels. There are 21 control channels in Band A and 21 control channels in Band B. All control channels carry data information, which is transmitted by means of FSK modulation with ±8-kHz frequency deviation.

The FOCC is transmitted from the base station to the mobile for paging, channel assignment, overhead, etc. Data transmission over the FOCC is based on encoding 28 control bits into a (40,28,5) BCH code. The encoded word is then transmitted by means of FSK modulation with ±8-kHz discrete level frequency deviations to represent 1 and 0. On the receiving side the RF signal is demodulated, decoded, and the original data finally recovered. Because this is a radio channel, the recovered data are impaired by noise, interference, and fading. As a result, the information is subject to degradation, causing a reduction in the FOCC capacity.

The RECC is transmitted by the mobile to the base station to originate a call. It is also known as an access channel, which is used by the mobile to access a land telephone or a mobile telephone. Data transmission over the RECC is based on encoding 36 control bits into a (48,36,5) BCH code. This means that if the decoded word encounters more than two errors due to noise, interference, and fading, an alarm will be generated and the word will be declared invalid, thus reducing the RECC capacity.

Both the FOCC and RECC are full duplex and operate in a coordinated manner. The objective of this section is to examine the control channel while operating in the presence of interference and fading and to determine its call handling capacity.

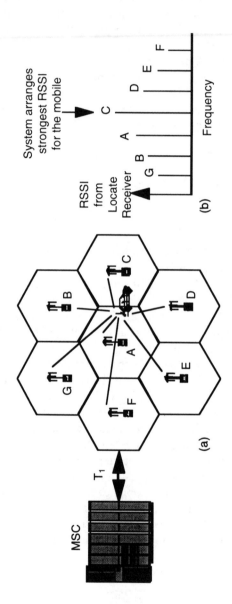

Figure 3.7 (a) Mobile in base A, moving toward base C. Its signal strength declines and becomes a hand-off candidate. (b) All located receivers monitor the signal strength of the mobile. MSC arranges the strongest RSSI for the mobile.

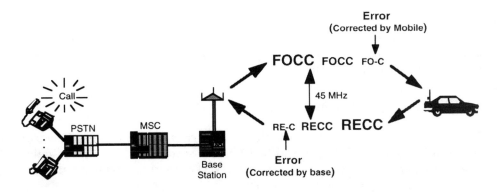

Figure 3.8 Symbolic representation of the control channel showing forward and reverse path.

3.5.2 Control Channel Modulation and Demodulation Techniques

Data transmission over the control channel is based on FSK modulation with ±8-kHz discrete level frequency deviations as shown in Figure 3.9. In this technique, the carrier assumes one frequency for binary symbol 1 and another frequency for binary symbol 0. This is represented as:

$$S_1(t) = A \cos(\omega_o + \Delta\omega_o)t \text{ for binary 1}$$
$$S_2(t) = A \cos(\omega_o - \Delta\omega_o t) \text{ for binary 0} \qquad (3.3)$$

where ω_o is the unmodulated carrier frequency. Thus, upon application of the input binary signal, the carrier shifts back and forth from the nominal frequency f_o by an

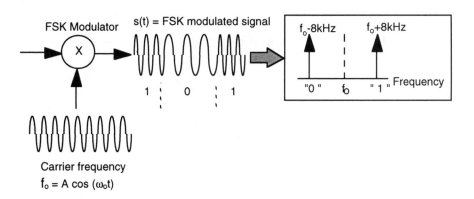

Figure 3.9 FSK modulation used in cellular control channel. Frequency deviation due to FSK modulation is ±8 kHz, where $f_0 + 8$ kHz represents 1 and $f_o - 8$ kHz represents 0.

amount $\pm\Delta f_o$ where Δf_o = 8 kHz. An infinite number of sidebands is generated during this process, as shown in Figure 3.10. These sidebands are generally attenuated by means of a filter. However, a residue is always present; these residual signals give rise to adjacent channel interference.

3.5.3 Forward Control Channel Description

The FOCC (Figure 3.11) is a continuous 10-Kbps data stream sent from the base station to the mobile. It begins with a 10-bit sync word for bit synchronization followed by an 11-bit sync word for frame synchronization. Following the sync word are three unique information streams:

1. Stream A = MIN (if LSB = 0).
2. Stream B = MIN (if LSB = 1).
3. BIS = busy/idle stream; busy = 0, idle = 1.

The BIS bit is imbedded into the A-B data stream (one busy/idle bit every 10 data bits). It indicates whether the RECC is occupied or not. The frame length is 42.1 ms where the BIS bit repeats at 1 Kbps. Figure 3.12 shows the structure of a control channel message and the following list describes its features:

T_1T_2 = 00	One word is sent
T_1T_2 = 10	Multiple words are sent
DCC = 00, 10, 01, 11	Distinguishes cochannel control channels
SCC =	00 = 5970-Hz SAT color code
	01 = 6000 Hz
	10 = 6030 Hz

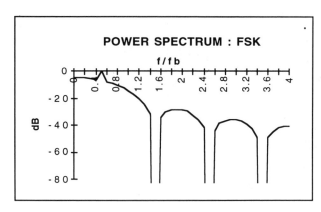

Figure 3.10 Normalized FSK power spectrum. Side bands are attenuated by means of a filter.

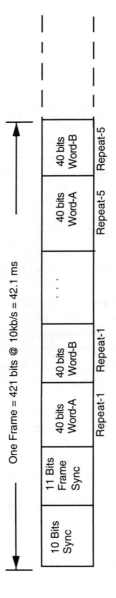

Figure 3.11 Forward control channel frame structure.

1. MOBILE STATION CONTROL MESSAGE

A. Abbreviated Address Word (40 Bits).

2	2	24	12
T₁T₂	DCC	MIN-1	P

B. Extended Address Word (40 bits)

2	2	10	1	5	3	5	12
T₁T₂ 1 0	SCC=11	MIN-2	RSVD=0	Local	OrdrQ	Order	P

C. Extended Address Word (40 bits)

2	2	10	3	11	12
T₁T₂ 1 0	SCC≠11	MIN-2	VMAC	CHAN.	P

Figure 3.12 Control channel message structure.

MIN1	Mobile identification number; seven-digit telephone number (24 bits)
MIN2	10-bit area code
VMAC	Voice mobile attenuation code (control mobile power)
Chan.	Channel assignment code
P	Parity bits

Order, OrdQ (order qualification), describes 11 different functions, including alert, release, stop alert, audit, registration, intercept, maintenance, send called address, direct retry status.

Figure 3.13 shows the structure of an overhead message and the following list describes its features:

- OHD/CMAC is a 3-bit code used to derive several functions as described earlier. The overhead message transmission rate is 1 sec.
- REGID (Registration ID) consists of a 20-bit word that represents the registration ID field.

3.5.4 Reverse Control Channel Description

The RECC (Figure 3.14) is a wideband 10-Kbps data stream sent from the mobile unit to the base station. It begins with 48 sync bits for bit synchronization followed by five 48-bit words (A, B, C, D, E), each repeated five times for redundancy.

2. OVERHEAD MESSAGE

2	2	20	1	3	12
T₁T₂	DCC	REGID	END	OHD/CMAC	P

OHD CODE	FUNCTION
000	Registration ID
001	Control Filler (Also CMAC)
010	Reserved
011	Reserved
100	Global Action
101	Reserved
110	Word-1 of System Parameter Msg.
111	Word-2 of System Parameter Msg.

Figure 3.13 Overhead message structure.

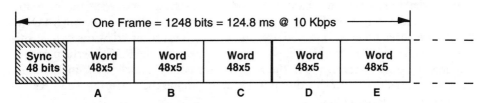

Figure 3.14 Reverse control channel frame structure.

Data transmission is based on BCH encoding. The encoded bit sequence is then transmitted by means of FSK modulation. Because the 48-bit information word is repeated five times for redundancy, the accept/reject decision is based on three good words out of five. This selection process is carried out by the base station radio. The frame length is 124.8 ms having 1248 bits/frame and the bit rate is 10 Kbps. A brief description of the RECC message structure is given in Table 3.2 and the types of messages that are sent over the RECC are shown in Table 3.3.

3.6 CONTROL CHANNEL CAPACITY

Control channel capacity [3,4] is a major concern in cellular communication because there is only one control channel per sector and three control channels per cell. Therefore, we need to identify the factors that limit the capacity.

To proceed with this exercise, let us consider the system model shown in Figure 3.15 where noise is introduced in both the forward and reverse control channels.

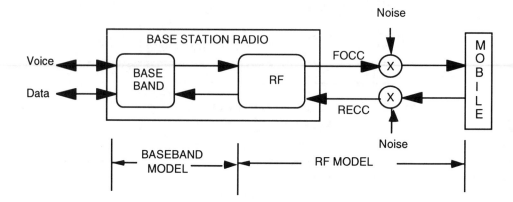

Figure 3.15 Control channel evaluation model.

The channel is partitioned into two functional blocks: RF and base band. It is assumed that the decision mechanism is resident in the base band where the incoming impaired data through the RECC are processed. The outgoing data through FOCC are also impaired by noise, interference, and fading; it is processed by the mobile and is not considered in this analysis. It is further assumed that the FOCC performance is degraded by data impairments in the RECC since both channels work in a coordinated manner. Because of this dependency, we examine the performance of RECC first, followed by the FOCC.

3.6.1 RECC Capacity

The RECC frame (Figure 3.14) is based on a 48-bit sync word, followed by five 48-bit words, each repeated five times for redundancy. Assuming that the pass/fail decision is based on detection of three correct words out of five in any order, we obtain:

$$P_d = \sum_{i=0}^{2} \binom{5}{i} \mathrm{WER}^i (1 - \mathrm{WER})^{5-i} \tag{3.4}$$

where P_d is the probability that three out of five words are correctly detected and WER is the word error rate.

There are 48 bits per encoded word and each word detects two errors (EIA standard). Therefore, the WER is given by:

$$\mathrm{WER} = \sum_{j=2+1}^{N} \binom{N}{j} \mathrm{BER}^j (1 - \mathrm{BER})^{N-j} \tag{3.5}$$

where $N = 48$; $j = 3, 4, \ldots, 48$; and BER is the bit error rate. Being a cellular radio channel, the BER depends on the carrier-to-interference ratio (C/I) and Rayleigh fading. This can be computed as [4,5]:

$$\text{BER} \approx \left[\frac{1}{2 + C/I} \right] \quad \text{(Thermal noise neglected)} \tag{3.6}$$

$$C/I \approx \left(\frac{1}{N-1} \right)\left(\frac{D}{R} \right)^{\gamma} \tag{3.7}$$

where

$D/R = \sqrt{3N}$

R = radius of the cell

D = repeat distance

N = frequency plan (N = 4, 7, 9, etc.)

γ = propagation constant ($\gamma = 2$ in free space, $\gamma > 2$ elsewhere).

According to EIA specifications, the busy/idle bit must remain busy for at least 30 ms after the reception of the last word from the mobile. This is to accommodate false detection of words due to interference and fading. If a mobile cannot complete the call within RECC frame time $T_R + 30$ ms, the mobile is given additional time to complete the call. This process involves a serial search algorithm that continues the search until the correct word is obtained. The average elapsed time to reach a good agreement is given by:

$$
\begin{aligned}
T_{\text{acq}} &= (T_R + \Delta t) + \{\Delta t(P_F) + 2\Delta t(P_F)^2 + \ldots\} \\
&= (T_R + \Delta t) + \Delta t(P_F)\sum_{n=1}^{\infty} n(P_F)^{n-1} \\
&= T_R + \Delta t + \Delta t(P_F)/(1 - P_F)^2
\end{aligned}
\tag{3.8}
$$

where

T_{acq} = average acquisition time

T_R = RECC frame time, ms

$\Delta t = 30$ ms

P_F = false detection probability.

The term $\Delta t + \Delta t(P_F)/(1 - P_F)^2$ is the BIS (busy idle status) delay, which is a function of C/I. The BIS bit (busy = 0, idle = 1) is inserted into the FOCC data stream, one BIS bit every 10 bits (at 1 Kbps). It indicates whether the RECC is occupied or not. The effective capacity then becomes [3]:

$$\text{Capacity} = \frac{1}{T_{\text{acq}}} = \frac{1}{T_R + \Delta t + \Delta t(1 - P_d)/P_d^2} \tag{3.9}$$

where

$$P_d = 1 - P_F = \text{correct detection probability}$$
$$\Delta t + \Delta t(1 - P_d)/P_d^2 = \text{BIS delay}$$
$$T_R = 124.8 \text{ ms (constant)}$$
$$\Delta t = 30 \text{ ms (minimum)}.$$

Equation (3.9) is plotted in Figure 3.16 as a function of C/I, which shows that the capacity is insensitive to C/I for $C/I \geq 14$ dB; this is due to error control coding. The capacity degrades slowly for 12 dB $< C/I < 14$ dB and degrades rapidly for $C/I \leq 12$ dB. In the absence of interference, $P_d = 1$ and the capacity becomes 23,256/hr at a BIS delay of 30 ms and 11,880/hr at a BIS delay of 175 ms.

We now prove that the preceding result is correct as follows: The busy/idle bit remains busy for 30 ms (minimum) after the reception of the last word from the

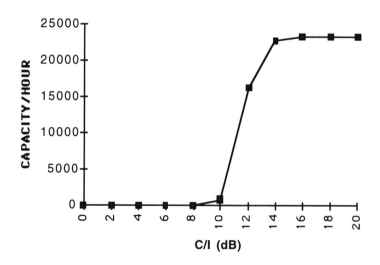

Figure 3.16 RECC capacity as a function of C/I.

mobile. Therefore the minimum occupancy time = 124.8 ms + 30 ms = 154.8 ms. In a fading environment, the busy/idle bit remains busy for 175 ms (maximum) after the reception of the last word from the mobile. Therefore, the maximum occupancy time = 124.8 ms + 175 ms = 299.8 ms. Therefore, the acquisition time will be given by:

$$T_{acq} = 154.8 \text{ ms (minimum)}$$
$$= 299.8 \text{ ms (maximum)}$$

yielding the RECC capacity as

$$\text{RECC capacity} = 1/T_{acq} = 3.33/\text{sec } (11988/\text{hr) (minimum)}$$
$$= 6.46/\text{sec } (23256/\text{hr) (maximum)}$$

which agrees with Figure 3.19.

3.6.2 FOCC Paging Capacity

Paging is interrupted during channel assignment, once every T_{acq} [3]. Thus, page interruptions due to channel assignment are given by:

$$\text{Page interruptions} = T_F/T_{acq} \qquad (3.10)$$

where

$$T_F = 42.1 \text{ ms}$$
$$T_{acq} = T_R + \text{BIS delay}$$
$$\text{BIS delay} = \text{busy/idle stream delay}$$
$$T_R = 124.8 \text{ ms.}$$

With a BIS delay of 30 ms (minimum) and 175 ms (maximum), the page interruption due to channel assignment will vary between 14% and 27% of the time. According to EIA standard, the overhead message repeats every 0.8 sec. This translates into a page interruption of approximately 5% of the time. The combined page interruption, therefore, varies between 19% and 32% of the time. Thus, the FOCC will be occupied by paging between 81% and 62% of the time.

The capacity can be expressed as:

$$\text{Number of page originations} = \frac{1}{T_F}\left(1 - \frac{T_F}{T_{acq}}\right)$$

$$1 - \frac{T_F}{T_{acq}} = \text{page occupancy time} \tag{3.11}$$

$$T_{acq} = T_R + \Delta t + \Delta t(1 - P_d)/P_d^2$$

and P_d is the detection probability, which is a function of C/I. Equation (3.11) is plotted in Figure 3.17 as a function of C/I.

For $C/I \leq 10$ dB, the RECC is overcome by interference, no channel assignment takes place, page interruption is zero, and hence the number of page originations is at maximum. As C/I increases, the RECC opens, the rate of channel assignments increases, and page interruption increases, thus reducing the paging capacity. Therefore, the performance of the FOCC is inversely proportional to the performance of the RECC.

3.6.3 Summary of Control Channel Capacity

A theoretical performance analysis of the cellular control channel is provided. Both the FOCC and RECC were examined and their capacities evaluated as a function of C/I. It is shown that the CCC is a function of C/I. The RECC capacity varies

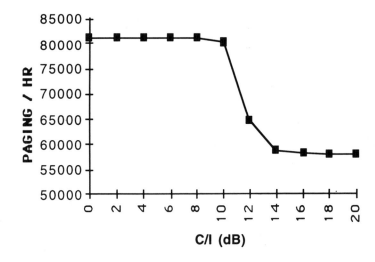

Figure 3.17 FOCC paging capacity as a function of C/I.

between 11,880/hr (3.3/sec) to a maximum of 23,256/hr (6.46 sec). The number of page originations over the FOCC varies between 52,920/hr (14.7/sec) and 69,264/hr (19.24/sec).

The analysis concludes that as the system traffic grows, a single control channel/OMNI cell would be overloaded, affecting services to many cellular subscribers. A separate control channel would be needed to mitigate this problem.

3.7 AMPS VOICE CHANNEL

The AMPS voice channel is composed of a forward voice channel (FOVCH) and a reverse voice channel (REVCH) over which voice and signaling transmission takes place between the base station and mobile units (Figure 3.18). Channel separation between the FOVCH and REVCH is 45 MHz. There are 395 voice channels and 21 control channels per band. All voice channels carry analog voice, signaling, and data information. It is important to note that voice and signaling take place in the analog domain (FM) and data transmission takes place in the digital domain (FSK) during hand-off. During this period ($\approx$100 to 200 ms) the voice is muted and the channel becomes a digital channel (FSK modulation), similar to the control channel.

Each voice channel pair supports a single conversation at a time. Four different types of signals are transmitted over the voice channel during the course of a cellular call:

1. Voice signals;
2. Supervisory audio tone (SAT);
3. Signaling tone (ST);
4. Data.

SAT and ST are imbedded into the voice (multitone FM). Voice is muted during hand-off and becomes an FSK-modulated channel for approximately 200 ms during hand-off completion. A brief description of these signals along with the method of transmission is presented in the following sections.

3.7.1 AMPS Voice Signal Transmission

On the transmit side, the voice signals are first compressed by a 2:1 syllabic compressor and then modulated by an FM modulator with ±12-kHz frequency deviation. On the receiving side, the incoming signal is demodulated and decompressed as 1:2 to recover the voice. Preemphasis and deemphasis (high-pass and low-pass) circuits, having a 6 dB/octave frequency response, are also used to improve the performance of FM transmission.

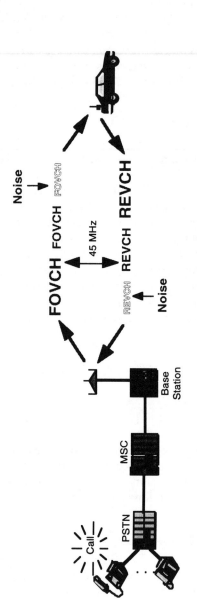

Figure 3.18 Symbolic representation of forward and reverse voice channel.

Frequency Modulation (FM)

In FM, the high-frequency carrier is varied in accordance with the low-frequency input signal as shown in Figure 3.19 where

$m(t)$ = low-frequency input modulating signal
$C(t)$ = high-frequency carrier signal
$s(t)$ = high-frequency modulated carrier signal.

In the process of shifting the carrier back and forth, additional sidebands are generated with the following frequencies:

$$(f_c \pm f_m), (f_c \pm 2f_m), \dots, \text{etc.} \qquad (3.12)$$

For a single tone modulation, each of the sidebands is separated from its neighbor by an amount equal to the frequency of the modulating signal. Figure 3.20 illustrates the mechanism before and after modulation. After modulation, power is taken from the carrier and distributed among the sidebands where the adjacent sidebands retains most of the power. Higher order sidebands are relatively weak and they are generally attenuated by a postmodulation filter. For these reasons, an adjacent channel can coexist with an acceptable level of C/I.

We further note that a low-intensity modulating signal takes less energy from the carrier and generates fewer sidebands. On the other hand, a high-intensity modulating signal takes more energy from the carrier and generates a large number of sidebands as shown in Figure 3.21. It follows that the FM transmission bandwidth depends on (1) the intensity of the modulating signal and (2) the frequency of the modulating signal. This led to the definition of the modulation index as follows:

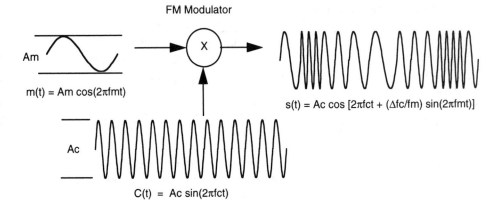

Figure 3.19 Frequency modulation (FM).

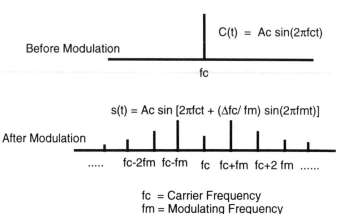

Figure 3.20 FM sidebands.

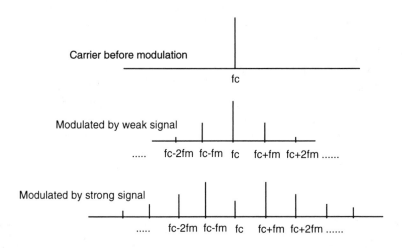

Figure 3.21 FM sideband power due to weak and strong modulating signal.

$$\text{Modulation index} = \frac{\text{Deviation of carrier frequency}}{\text{Modulating frequency}} \quad \text{or} \quad \mu = \frac{\Delta f_c}{f_m} \quad (3.13)$$

It follows that both frequency and intensity of the modulating voice signal have to be controlled in order to maintain the power spectrum within the 30-kHz transmission bandwidth. For these reasons, a 2:1 syllabic compander is used to control the intensity of the voice and preemphasis/deemphasis circuits (high-pass/low-pass filters) are used to control the frequency of the voice signal.

3.7.2 Supervisory Audio Tone (SAT)

The SAT is transmitted over the forward voice channel (base to mobile) and retransmitted by the mobile back to the base. Its purpose is to indicate the continuity of the conversation. Loss of SAT indicates call completion and hand-off. Only three SAT frequencies are available (5970, 6000, and 6030 Hz). Each SAT is assigned to a cluster of cells in such a way that the same SAT is not used in a reused channel. SAT frequencies distinguish cochannel sites. SAT features are summarized in the following list:

- Each base station is assigned a SAT tone.
- The SAT tone is transmitted over the voice channel.
- The mobile unit detects the SAT and transmits back the same SAT to the base station.
- SAT continues during the call.
- If SAT ≠ SCC (SAT color code), voice is muted.
- If SAT is invalid, the transmitter is turned off.
- If no valid SAT is received within 5 sec, the transmitter is turned off.

3.7.3 Signaling Tone (ST)

A 10-kHz ST is transmitted by the mobile over the reverse voice channel to acknowledge certain commands received from the base station. This is similar to the supervisory signaling tone used in conventional telephone networks. A 50-ms burst of ST indicates an acknowledgment of hand-off.

3.7.4 Data Transmission Over the Voice Channel

During hand-off, the frequency-modulated voice channel momentarily becomes an FSK-modulated digital channel, similar to a control channel. A 10-Kbps data transmission takes place between the base and the mobile for channel assignment and signaling. Voice tones, SAT, and ST are muted during this process. Once the channel assignment is complete, regular conversation resumes. About 200 ms worth of voice is muted during this process, which may be heard as a clicking noise during a conversation.

Voice Channel Structure During Hand-Off

During hand-off, the following steps occur:

- During the hand-off process, the voice channel is used for signaling (channel assignment data), as shown in Figure 3.22.
- Modulation: FSK with ±8-kHz deviation (similar to the control channel).

Forward Voice Channel During Handoff

101	11	40	37	11	40	
Bit Sync.	Word Sync	Word	Bit Sync	Word Sync.	Word	Repeated 11 times

Reverse Voice Channel During Handoff

100	11	40	37	11	40	
Bit Sync.	Word Sync	Word	Bit Sync.	Word Sync.	Word.	Repeated 5 times

Figure 3.22 AMPS voice channel frame structure during hand-off.

- 40-bit data is BCH encoded.
- Duration of the hand-off process is ≈200 ms.
- The word repeats 11 times to ensure proper reception of data because the mobile is in the hand-off region and the signal is weak.

3.8 AMPS LOCATE CHANNEL AND ITS FUNCTIONS

3.8.1 General Description

Each active voice channel is sampled by a scanning receiver known as a locate channel receiver, which is located at the base station (Figure 3.23). Upon receiving a command from the MSC, these receivers scan neighboring mobiles and measure the respective signal strengths, detect the presence of SAT, etc.

The detected signal strengths are reported to the MSC by each cell site while the MSC determines the location and the direction of travel of the mobile and directs it to tune to another frequency from an adjacent cell or sector. This hand-off process results in the locate receiver playing an important role in collecting a set of received signal strengths, under the command of the MSC. These signal strengths and the corresponding trends are then used by the MSC to determine the destination of the mobile for a possible hand-off to an adjacent cell or a sector. The absence of a SAT is an indication of hand-off or a call drop.

Once again, each OMNI cell site has at least one locate receiver, three locate receivers for three sector sites, and six locate receivers for six sector sites. These activities, performed by locate receivers, are also known as call supervision, which lasts for the full duration of the call. Therefore, a locate receiver is subject to capacity

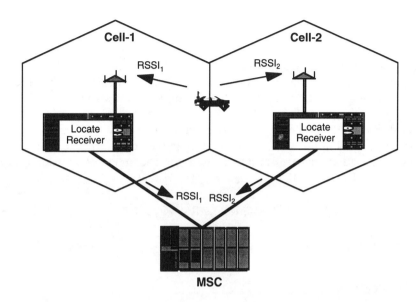

Figure 3.23 AMPS locate channel receiving mechanism, located in each cell or sector. The transmitter is disabled from each locate radio.

limitations. For instance, if the messaging due to hand-off exceeds the limitation of the locate receiver, the call will be dropped. An additional locate receiver would be needed to mitigate this problem.

3.8.2 Hand-Off Processing

Hand-off is a process of changing the frequency. Its primary purpose is to assign a new frequency when a mobile moves into a new cell or a sector. This is accomplished by setting a hand-off threshold, that is, if the received signal level is too low and reaches a predefined threshold, the system controller, namely, the mobile switching center, provides a stronger free channel (frequency) from an adjacent cell. Hand-off provides mobility, which means that a caller can move without interruption or call drop. A brief description of hand-off protocol was provided in Chapter 2.

The basic hand-off process can be broken down into several stages, including these:

- Measurement stage;
- Trigger stage;
- Screening stage;
- Selecting stage;
- Execution stage.

3.8.2.1 Measurement Stage

RSSI measurements are taken for each traffic and averaged by each locate receiver. Both long- and short-term averages are accumulated and reported to the MSC. Thus referring to Figure 3.24, we see that, as the mobile moves away from serving base A, its signal strength RSSIA declines and reaches a predefined threshold, HOTL (hand-off threshold). When the signal strength drops below the threshold, the base station reports this to the MSC, thus initiating the hand-off process. The MSC then advises all the adjacent base stations to monitor and report RSSI readings of the mobile. A set of RSSI readings,

$$\{RSSI_A, RSSI_B, RSSI_C, \ldots\} \tag{3.14}$$

is then received at the MSC for further processing, such as triggering and screening.

3.8.2.2 Trigger Stage

Triggering is an action that starts the hand-off process. This is initiated by the locate receiver after the mobiles RSSI drops below the HOTL. This is shown in Figure 3.25 where RSSI is plotted as a function of distance. The condition for triggering is

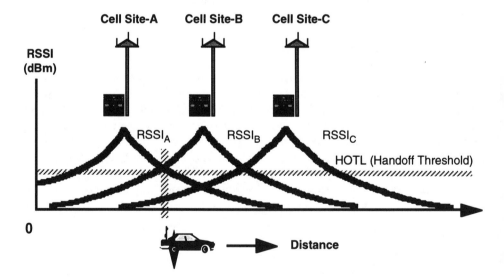

Figure 3.24 Received signal strengths as measured by locate receivers.

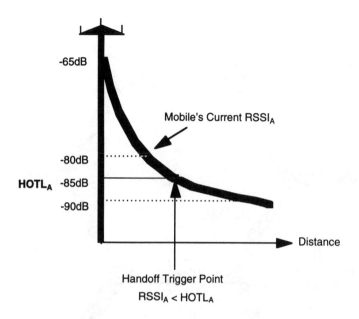

Figure 3.25 Basic concept of hand-off triggering.

$$RSSI_A \leq HOTL \qquad (3.15)$$

where the RSSI threshold HOTL is defined within the MSC during cell site design. This is a user-defined parameter that varies from cell site to cell site.

In a fully developed system, the triggering process is more complex because of multiple cell adjacency as shown in Figure 3.26. For example, the RSSI threshold for Cell-A may be defined with respect to six neighboring cells:

$$\{HOTL_{AB}, HOTL_{AC}, HOTL_{AD}, HOTL_{AE}, HOTL_{AF}, HOTL_{AG}\} \qquad (3.16)$$

where the RSSI trigger value for Cell A is the strongest of all, that is:

$$RSSI_A \text{ Trigger } (HOTL_A) = \qquad (3.17)$$
$$max\{HOTL_{AB}, HOTL_{AC}, HOTL_{AD}, HOTL_{AE}, HOTL_{AF}, HOTL_{AG}\}$$

This implies that if $RSSI_A$ drops below $HOTL_A$, locate requests will be sent to all adjacent cell sites meeting these criteria:

$$RSSI_A < HOLTLA\# \qquad (3.18)$$

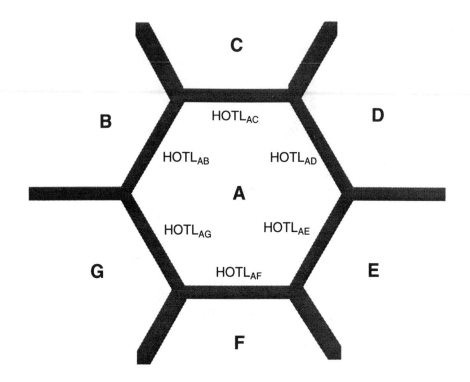

Figure 3.26 RSSI contour for hand-off triggering in a multicell environment.

but not to those cells meeting these criteria:

$$RSSI_A > HOLTLA\#$$ (3.19)

3.8.2.3 Screening Stage: Hysteresis

Screening is the process of qualifying a set of candidate cells from neighboring cells. It is generally based on signal strength measurement but also includes an additional factor such as *hysteresis,* which prevents hand-off back and forth, commonly known as *ping-pong.*

Hysteresis is a differential RSSI value applied during the screening process. The mechanism is illustrated in Figure 3.27 for two adjacent cells: Cell A and Cell B. Cell A is the serving cell for mobile A and Cell B is the serving cell for mobile B. Assuming that mobile A is moving toward Cell B, with hysteresis, it is not eligible for hand-off from Cell A to Cell B until the mobile's RSSI at Cell B is stronger than the one received at Cell A by an amount $Hyst_{AB}$. This is governed by the following logic:

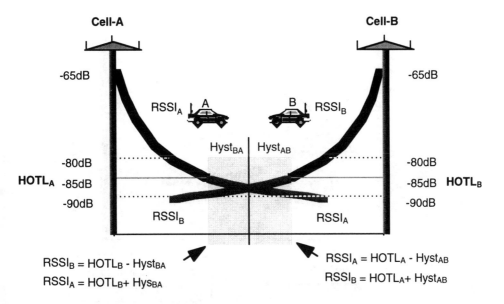

Figure 3.27 Hand-off screening process showing the use of hysteresis.

$$RSSI_B > HOTL_A + Hyst_{AB} \qquad (3.20)$$

which is essentially an artificial shift in $RSSI_A$, protecting the mobile from hand-off back to Cell A. This mechanism is graphically shown in Figure 3.28.

Similarly, if mobile B is moving toward Cell A, with hysteresis, it is not eligible for hand-off from Cell B to Cell A until the mobile's RSSI at Cell A is stronger than the one received at Cell B by an amount $Hyst_{BA}$, and the logic is

$$RSSI_A > HOTL_B + Hyst_{BA} \qquad (3.21)$$

The hysteresis value is measured in decibels, which varies from cell site to cell site, depending on the propagation environment. Typical values ranges from 3 to 5 dB. There is also a compromise involved when setting the hysteresis value. For example, too much hysteresis will result in no hand-off or "call dragging," and a small hysteresis will generate excessive locate channel activity due to ping-pong. The end result is excessive messaging, which must be processed by the MSC, resulting in an undesirable reduction of switch capacity.

3.8.2.4 Selection With Hysteresis

Selection is a process of choosing a cell from a set of candidate cells that survived the screening process. The logic that governs this process is given by:

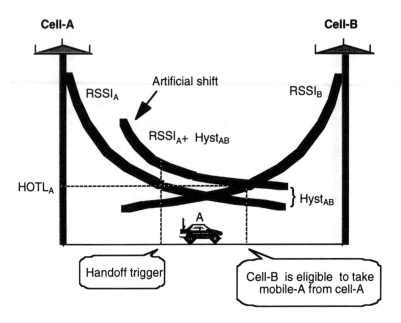

Figure 3.28 Graphic representation of an artificial shift of RSSI$_A$. Mobile A is served by Cell A.

$$\max\{HOTL_A + Hyst_{AB}, HOTL_A + Hyst_{AC}, \ldots\} \qquad (3.22)$$

which is fixed and does not consider irregular traffic distribution patterns. As a result, an unnecessary hand-off takes place into a congested cell, lasting for a short duration of time. A possible solution to this problem is *hand-off biasing*, which offers a deliberate preference over another cell. It can be used during the selection process to steer traffic to a noncongested cell and maintain uniformity of traffic distribution in the service area. The concept of biasing is presented in Section 3.8.2.5.

3.8.2.5 Selection With Hand-Off Biasing

Analogous to hysteresis, hand-off biasing is also a differential factor, used to select a target cell where a deliberate preference is given to one cell over another. This is an important tool that can be used to steer traffic from a busy cell to a nonbusy cell. It also helps to reduce unnecessary hand-offs. To illustrate the concept, we consider the scenario shown in Figure 3.29. Cell A is the serving base for mobile A, which is approaching Cell B and then Cell C. If the mobile is momentarily handed off to Cell B, it becomes a hand-off candidate again, this time for Cell C. This momentary hand-off to Cell B is an unnecessary activity that can be avoided by means of biasing.

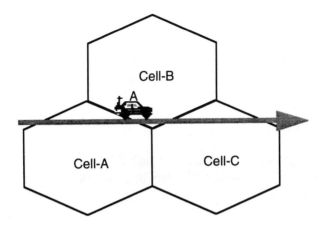

Figure 3.29 Before bias.

One possible solution to this problem would be to apply a positive bias to Cell C and make it artificially better than Cell B as shown in Figure 3.30(a). In this case the hand-off will take place directly from Cell A to Cell C. Alternately, we can apply a −ve bias to Cell B and make it artificially worse than Cell C as shown in Figure 3.30(b). In this case, the hand-off will take place directly from Cell A to Cell C. In both cases, the mobile will skip Cell B. Thus, biasing modifies the relative ranks of target cells. Logical statements for the selection process with bias are as follows:

$$\text{If the adjusted RSSI}_C > \text{Original RSSI}_C + \text{Bias}_{AC} \text{ (+ve Bias)} \qquad (3.23)$$

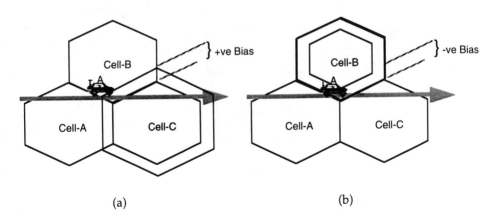

Figure 3.30 Hand-off bias: (a) +ve bias and (b) −ve bias.

then choose Cell C, or

$$\text{If the adjusted RSSI}_B > \text{Original RSSI}_B - \text{Bias}_{AB} \ (-\text{ve Bias}) \qquad (3.24)$$

then choose Cell C.

3.8.2.6 Execution Stage

The execution stage is a process of channel assignment from the final candidate cell and advice to the mobile to tune to the new frequency. Channel availability is also carried out during this process.

3.8.3 Locate Channel Capacity

There is only one locate receiver per cell and sector. Therefore, as the system traffic grows, a single locate receiver can become overloaded, affecting services to many subscribers. An additional locate receiver is needed to mitigate this problem. As such it is desirable to assess the capacity before provisioning a second locate receiver. To compute its capacity, we consider the conceptual model of a scanning locate receiver as shown in Figure 3.31. It consists of a locate receiver, driven by a frequency synthesizer. The function of the frequency synthesizer is to synthesize a set of frequencies, under the command received from the MSC.

Since the function of the locate receiver is to measure RSSI and detect SAT from several channels, the capacity can be computed as

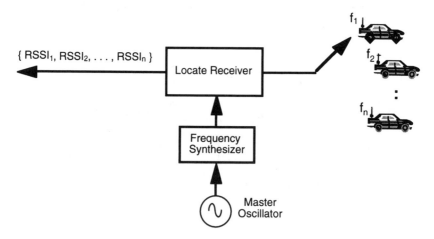

Figure 3.31 Illustration of locate receiver capacity.

$$\text{Locate channel capacity} \approx \frac{1}{T_{\text{L}} + T_{\text{RSSI}} + T_{\text{SAT}}} \qquad (3.25)$$

where T_{L} is the frequency lock time, T_{RSSI} is the RSSI measurement time, and T_{SAT} is the SAT measurement time. For example, if $T_{\text{L}} = 25$ ms, $T_{\text{RSSI}} = 50$ ms, and $T_{\text{SAT}} = 150$ ms, we obtain

$$\text{Locate channel capacity} = \frac{1}{25 \text{ ms} + 50 \text{ ms} + 150 \text{ ms}} = 4.44/\text{sec} \ (16{,}000/\text{hr})$$

References

[1] IS-54, "EIA Project Number 2215," 1989, pp. 3/18–3/47.

[2] "Advanced Mobile Phone Services," Special Issue, *Bell Syst. Tech. J.*, Vol. 58, January 1979.

[3] Faruque, S. M., "Cellular Control Channel Capacity," *Can. J. Electr. Comput. Eng.*, Vol. 19, No. 1, January 1994, pp. 13–15.

[4] Faruque, S. M., "Cellular Control Channel Performance in Noise, Interference and Fading," *Proc. IEEE Int. Conf. Selected Topics in Wireless Communications*, IEEE, 1992, pp. 328–331.

[5] Lee, William, C. Y., *Mobile Cellular Telecommunications Systems*, New York: McGraw-Hill Book Company, 1989.

[6] Mehrotra, Asha, *Cellular Radio Performance Engineering*, Norwood, MA: Artech House, 1994.

CHAPTER 4
▼▼▼

INTRODUCTION TO CDMA

4.1 INTRODUCTION

Code-division multiple access (CDMA) is a spread-spectrum communication system in which multiple users have access to the same frequency band. In DS-CDMA (direct sequence CDMA), spread spectrum refers to power spreading over a given transmission bandwidth. This is accomplished by spreading the base-band binary data by means of a high-speed pseudonoise (PN) code (called the *chip rate*). The composite high-speed data are then modulated and transmitted over the air. In the North American DS-CDMA standard (IS-95) [1], the rate of the PN sequence was chosen to be approximately 1.25 MHz (≈1.228 MHz) and the transmission bandwidth was chosen to be exactly 1.25 MHz. Ten different frequency bands are derived from the existing 12.5-MHz cellular carrier (A or B). Each of these bands supports 64 orthogonal codes known as Walsh codes, one Walsh code per user.

The objective of this chapter is to present the foundation of DS-CDMA and related topics associated with the IS-95 standard. It is primarily geared to beginners. However, those already exposed to a practical CDMA system but interested in understanding the concept of spectrum spreading/despreading techniques, PN codes, Walsh codes, process gain, soft/hard capacity, power control mechanisms, soft/hard hand-off, and transmitting/receiving structures may also use this material as a reference.

4.2 BASIC CONCEPT OF THE SPREAD SPECTRUM

Understanding of CDMA begins with the basic concept of spectrum and the process of spectrum spreading. Here, the term *spectrum* refers to the power spectrum associated with the base-band signal and the term *spread spectrum* refers to the spreading of the power spectrum of the base-band signal over a given bandwidth. This concept is briefly discussed in the next section.

4.2.1 Spectrum

We begin our review by considering a simple discrete time circuit as shown in Figure 4.1(a) that is loaded by a resistor R and driven by a nonperiodic discrete time signal having the following boundary conditions:

$$V(t) = V \quad < 0 < t < T$$
$$= 0 \quad \text{elsewhere} \tag{4.1}$$

This signal is also known as NRZ (nonreturn to zero) data, which is generally used in digital radio. Now, our goal is to determine the frequency content of this signal and then to evaluate the power spectrum associated with this signal.

To determine the frequency and power spectrum of the signal described in (4.1), we apply the Fourier transform:

$$S(\omega) = \int_0^T V \cdot e^{-j\omega t} dt$$

$$= \left(2\frac{V}{\omega}\right) \sin(\omega T/2)$$

$$= VT\left[\frac{\sin(\omega T/2)}{\omega T/2}\right] \tag{4.2}$$

which reveals that an NRZ datum is composed of an infinite number of harmonically related sinusoidal waves having different amplitudes as shown in Figure 4.1(b). Therefore, the power dissipated by the load resistance R will be due to all the sinusoidal components, which can be determined as:

$$P(\omega) = \left(\frac{1}{T}\right)|S(\omega)|^2$$

$$= V^2 T\left[\frac{\sin(\omega T/2)}{\omega T/2}\right]^2 \tag{4.3}$$

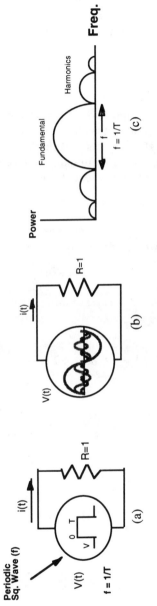

Figure 4.1 (a) A discrete time circuit. (b) Equivalent circuit. (c) Power spectrum.

Figure 4.1(c) shows the familiar power spectrum having a main lobe corresponding to the fundamental component of the frequency and infinite number of sidelobes corresponding to the harmonic components. We also note that most of the power is retained by the main lobe whose bandwidth (BW) is given by BW = 1/T, where T is the bit duration.

4.2.2 Spectrum Spreading

Spectrum spreading can be accomplished simply by increasing the frequency of the discrete time signal. Thus we consider a waveform with an amplitude V and frequency f ($f = 1/T$) and then increase the frequency of the same waveform by a factor of n, i.e., T is now reduced by n. A pair of boundary conditions describing this situation is given in (4.4) and the corresponding waveform is shown in Figure 4.2(a).

$$
\begin{aligned}
V(t) &= V &< 0 < t < T \\
&= 0 &\text{elsewhere} \\
V(t) &= V &< 0 < t < T/n \\
&= 0 &\text{elsewhere}
\end{aligned}
\tag{4.4}
$$

Applying the Fourier transform in (4.4) we obtain the following spectral components:

For $0 < t < T$

$$
S(\omega) = \int_0^T V \cdot e^{-j\omega t} dt = VT\left[\frac{\sin(\omega T/2)}{\omega T/2}\right]
\tag{4.5}
$$

$$
P(\omega) = \left(\frac{1}{T}\right)|S(\omega)|^2 = V^2 T\left[\frac{\sin(\omega T/2)}{\omega T/2}\right]^2
\tag{4.6}
$$

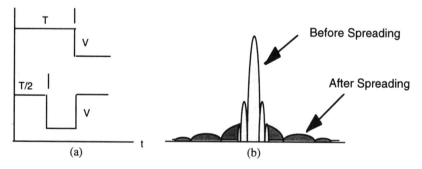

Figure 4.2 (a) Representation of a discrete time signal having an amplitude of V and two different frequencies, f and $2f$, where $f = 1/T$. (b) The corresponding power spectrum.

For $0 < t < T/n$

$$S(\omega) = \int_0^T V \cdot e^{-j\omega t} dt = VT\left[\frac{\sin(\omega T/2n)}{\omega T/2n}\right]$$

$$P(\omega) = \left(\frac{1}{T}\right)|S(\omega)|^2 = V^2 T\left[\frac{\sin(\omega T/2n)}{\omega T/2n}\right]^2 \tag{4.7}$$

Figure 4.2(b) shows the power spectrum for $n = 1$ and $n = 2$.

We now turn our attention to the energy delivered to the load between time $t = 0$ and $t = T$. This is given by the total area under the curve (Figure 4.2(b)),

$$E(t) = \int_0^T P(\omega)dt = \left(\frac{1}{T}\right)\int_0^T |S(\omega)|^2 dt = \text{Constant} \tag{4.8}$$

which means that the total energy under the power spectrum curve remains the same after spreading. It follows that if the spreading bandwidth is sufficiently high, the amplitude of the signal would be reduced accordingly. This is known as process gain and is described in the next section.

4.2.3 Process Gain

Process gain is due to spectrum spreading, defined as:

$$G_s = 10 \log\left(\frac{\text{BW}}{R_b}\right) \tag{4.9}$$

where G_s is the process gain, BW is the transmission bandwidth, and R_b is the bit rate. For example, if BW = 30 kHz, R_b = 10 kHz, then G_s = 10 log(30/10) = 4.77 dB. Now if we increase the bandwidth to 1.25 MHz, the process gain would be G_s = 10 log(1,250,000/10) = 20.97 dB, providing an additional 20.97 − 4.77 = 16.20 dB margin to suppress interference. In a multiple-access environment, this margin is reduced by 3 dB for 2 users, 10 dB for 10 users, etc. The thermal noise, which is not affected by the process gain, is also an unavoidable source of interference.

However, note that process gain alone is not responsible for overall system performance and capacity. Other factors such as error control coding, cell site deployment, antenna directivity and downtilt, sectorization, and frequency reuse play important roles in enhancing the overall performance and capacity of the CDMA system.

4.2.4 Summary of Findings

From the above discussions we find the following:

- A discrete time signal is composed of an infinite number of harmonically related sinusoidal waves (spectrum).
- The associated power is due to an infinite number of sinusoidal components.
- Energy is independent of frequency (conservation of energy).
- The bandwidth of the power spectrum is proportional to the frequency of the base-band signal (spectrum spreading).
- Since the energy is constant, the amplitude of the signal after spreading will be reduced. If the bandwidth is enough, the amplitude will be close to noise level.
- This process is known as *spread spectrum,* resulting in a process gain, defined as $G_s = 10 \log(BW/R_b)$, where BW is the bandwidth and R_b is the bit rate.
- Therefore, *spread spectrum* means spreading *the power spectrum over a predetermined band.*

4.3 DIRECT SEQUENCE SPREAD SPECTRUM

Direct sequence (DS) spread spectrum is based on directly spreading and despreading the base-band data by means of a PN sequence. The methodology is briefly presented in the following subsections.

4.3.1 DS Spectrum Spreading

DS spectrum spreading is accomplished by means of a two-input exclusive-OR gate (Figure 4.3(a)) where A is low-speed NRZ data and B is a high-speed PN sequence. The exclusive-OR gate (MOD2 adder) obeys the following logic: If A and B are identical, then C = 0; otherwise, C = 1. In Boolean expression this is given by

$$C = A\overline{B} + \overline{A}B \tag{4.10}$$

The corresponding truth table and the associated waveform are given in Figures 4.3(b) and (c), respectively. The power spectrum is also shown in the inset of Figure 4.3(a) as an illustration. Here, the input signal A is low-speed NRZ data having a narrow power spectrum. The second input is a high-speed signal B, generally a PN code, which has a wider power spectrum. The composite signal C at the output has the same transition rate as the PN sequence (the B signal), because of its wideband power spectrum, but a lower amplitude because the total energy is constant.

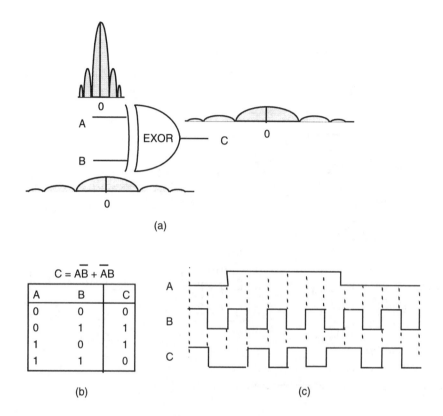

Figure 4.3 (a) An exclusive-OR gate. (b) Truth table. (c) Waveforms.

4.3.2 DS Spectrum Despreading

DS spectrum despreading is a process of data recovery from the composite spread-spectrum signal. This is accomplished by means of another exclusive-OR gate shown in Figure 4.4, where the composite data C is applied to one input and an identical PN sequence is applied to the second input. The output "Y" is a decomposed signal, which is the original NRZ data "A." This recovery process only works when the PN sequence (the B signal) is identical for both spreading and despreading; otherwise, the desired signal will never be recovered. We examine this process by means of the following example.

As an example we consider the circuit of Figure 4.5, where low-speed NRZ data (A) is MOD2 added with a high-speed PN sequence (B). The composite spread-spectrum signal (C) is MOD2 added once again by means of the same PN code. Our objective is to prove that the final output $Y = A$, if the PN sequences are identical.

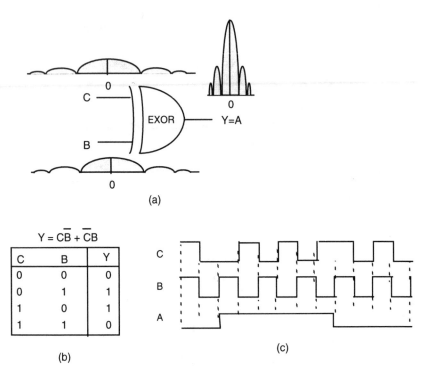

(a)

$Y = C\bar{B} + \bar{C}B$

C	B	Y
0	0	0
0	1	1
1	0	1
1	1	0

(b)

(c)

Figure 4.4 (a) An exclusive-OR gate. (b) Truth table. (c) Waveforms.

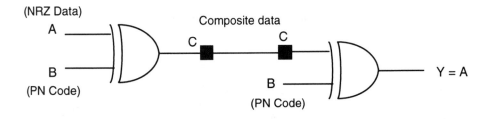

Figure 4.5 A circuit to illustrate spectrum spreading and despreading.

Using the Boolean expression, we obtain

$$C = A\bar{B} + \bar{A}B$$

$$Y = C\bar{B} + \bar{C}B \quad \text{or} \quad Y = (A\bar{B} + \bar{A}B)\bar{B} + \overline{(A\bar{B} + \bar{A}B)}B \qquad (4.11)$$

with

$$(\overline{A\overline{B} + \overline{A}B}) = (\overline{A} + B) \cdot (A + \overline{B})$$
$$A\overline{A} = B\overline{B} = 0,$$
$$\overline{B}B = \overline{B}$$
$$(B + \overline{B}) = 1$$

which yields

$$Y = A \qquad\qquad (4.12)$$

where both PN codes are identical; otherwise, $Y \neq A$.

4.4 DS SPREADING AND DESPREADING IN THE PRESENCE OF INTERFERENCE

One of the major advantages of the spread-spectrum signal is its tolerance to interference. We examine this feature by introducing an interferer as shown in Figure 4.6.
   Applying the Boolean algebra, we obtain

$$C = A\overline{B} + \overline{A}B \qquad\qquad (4.13)$$

$$Y = (C\overline{B} + \overline{C}B) + (n\overline{B} + \overline{n}B) \quad \text{or} \quad Y = (A\overline{B} + \overline{A}B)\overline{B} + \overline{(A\overline{B} + \overline{A}B)}B + (n\overline{B} + \overline{n}B)$$

which reduces to

$$Y = A + (n\overline{B} + \overline{n}B) \qquad\qquad (4.14)$$

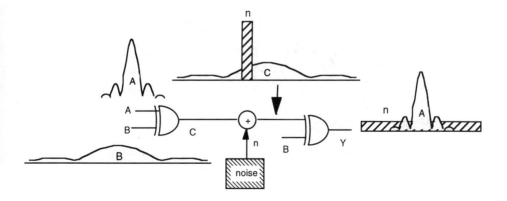

Figure 4.6 Spreading and despreading within an interference environment.

where A is the desired signal, recovered due to despreading, and the second term is the interference component, spread over the entire band of the high-speed PN code (B). Since the energy is conserved, the magnitude of the interference power is reduced proportionately, referred to as *process loss*. Therefore, all users, other than the desired signal, appears as noise. This is conceptually shown in Figure 4.7 where the desired signal is recovered by despreading; all other users remain spread over the entire transmission band.

Figure 4.7 implies that there is a *process loss* for the interferer due to spreading and there is a *process gain* due to despreading of the desired signal. For k users, this loss can be defined as

$$\text{Process loss} = 10 \log(k) \tag{4.15}$$

The overall system gain can be defined as

$$\begin{aligned}
\text{CDMA gain} &= \text{Process gain} - \text{Process loss due to } k \text{ users} \\
&= 10 \log(\text{BW}/R_b) - 10 \log(k) \\
&= 10 \log(\text{BW}/kR_b) \tag{4.16}
\end{aligned}$$

where k is the total number of users having access to the same CDMA band. Note that the spreading and despreading technique applies to other users and does not apply to the thermal noise.

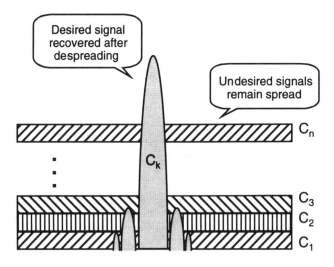

Figure 4.7 Illustration of signal recovery from several spread-spectrum signals. Each signal is spread by means of a unique code.

4.5 DS SPREAD-SPECTRUM MODULATION AND DEMODULATION TECHNIQUES

Spread-spectrum modulation is a process of modulating the spread-spectrum base-band signal by means of a suitable modulator. This is accomplished as a combination of a MOD2 adder (exclusive-OR gate) and a high-speed digital modulator as shown in Figure 4.8. The speed of the modulator is determined by the PN code rate. In North American CDMA, the rate of the PN code is specified as 1.2288 Mbps. Therefore the information rate at the output of the MOD2 adder is also at 1.2288 Mbps in which the NRZ data are imbedded. The output of the modulator is the modulated intermediate frequency (IF) signal.

Spread-spectrum demodulation is a reverse process, as shown in Figure 4.8. The spread-spectrum IF signal is first demodulated to obtain the composite spread-spectrum data. The composite data are then MOD2 added with the same PN code to recover the original NRZ data.

4.6 DS SPREAD-SPECTRUM RADIO

The basic functional block of a spread-spectrum radio is based on a transmitter, a receiver, and a timing generator as shown in Figure 4.9. Numerous other components such as error control coding, base-band filters, postmodulation filters, and PN generators are integral parts of the system, but are not shown in this figure. We now briefly describe the operation of the radio.

Transmitter

- The input NRZ data are MOD2 added with a high-speed PN code to obtain high-speed composite data. This PN code is unique and is not reused elsewhere in the same service area (cell).
- The high-speed composite data are then modulated by means of a high-speed PSK modulator to obtain the IF signal.
- Next, the IF signal is upconverted by means of a mixer to obtain the desired RF signal.
- Finally, the RF signal is amplified and transmitted by means of a suitable antenna.

Receiver

On the receiving side, the operation is a reverse process:

- The incoming RF signal is amplified by means of a low-noise amplifier (LNA) and then downconverted by means of a mixer to obtain the IF signal.

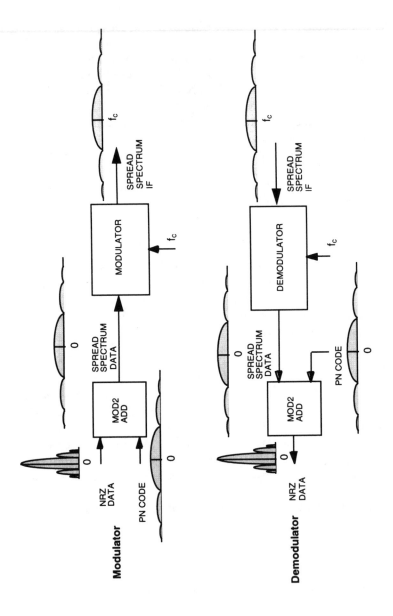

Figure 4.8 Spread-spectrum modulation and demodulation technique.

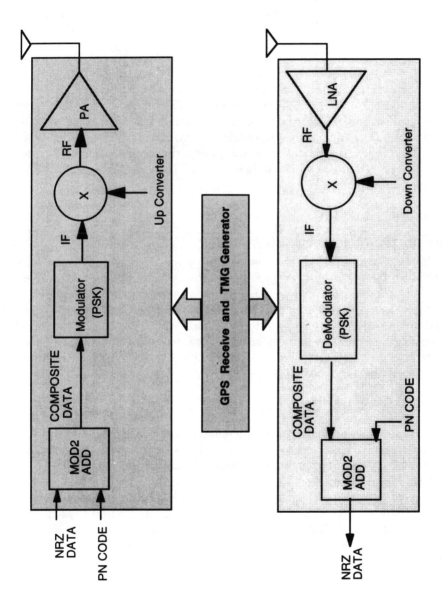

Figure 4.9 The conceptual block diagram of spread-spectrum radio.

- Next, the IF signal is demodulated to obtain the composite high-speed data where the desired traffic data are imbedded.
- Finally, the composite high-speed data are MOD2 added with the same PN code to recover the desired traffic.

Timing Generator

One of the crucial components in any spread-spectrum radio is the timing generator. It is based on a GPS receiver and timing circuitry. The function of the GPS receiver is to recover accurate timing from geostationary satellites, while the timing circuitry generates various timing waveforms and delivers them to various functional blocks of the radio. All base station radios recover timing from a single GPS receiver, while all mobiles recover timing from the base station. Therefore, mobile units do not have GPS receivers.

4.7 WHAT IS DS-CDMA?

CDMA is a branch of the multiple-access radio communication process in which multiple users have access to the same system, using the same frequency. This is accomplished by means of an m-bit PN generator, which provides $2^m - 1$ different codes. Out of these codes, only m codes, known as *orthogonal codes,* are derived and assigned to m users, one code per user. The function of the PN code is to spread the traffic data over the entire transmission band while uniquely identifying each user. Because the spreading and despreading are accomplished by direct application of a PN sequence, the overall process is described as direct sequence code-division multiple access or simply DS-CDMA.

4.7.1 Reverse Link DS-CDMA

As an illustration, we present a conceptual model of a reverse link DS-CDMA system, providing access to k mobiles: $M_1, M_2, \ldots, M_k$, using the same carrier frequency f_c, shown in Figure 4.10. Each mobile is assigned a unique PN code: $PN_1, PN_2, \ldots, PN_k$ where PN_1 is assigned to M_1, PN_2 to M_2, and so on. The CDMA base station is assumed to be a multiple-access point where all the propagated spread-spectrum signals arrive at random. It is the responsibility of the base station to identify each piece of traffic uniquely by means of an array of MOD2 adders, biased with the respective PN codes. Each MOD2 adder then despreads one of k signals, which is the desired traffic.

4.7.2 Forward Link DS-CDMA

The forward link DS-CDMA process is described in Figure 4.11. The incoming traffic from the T_1 link is spread by means of an array of MOD2 adders, biased

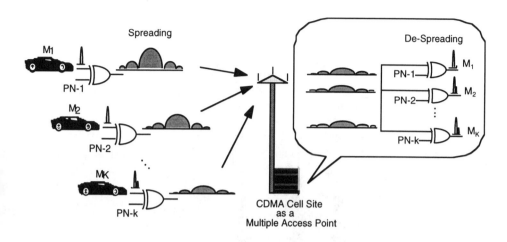

Figure 4.10 Conceptual representation of reverse link DS-CDMA.

with the respective PN codes (PN − 1, PN − 2, . . . , PN − k). Each spread-spectrum signal is then modulated, upconverted, and finally transmitted. These signals are received by all the mobiles in the service area and MOD2 added by the respective PN code to recover the desired traffic.

4.8 PN SEQUENCE

The pseudorandom noise (PN) sequence is extensively used in digital communication systems for data scrambling due to its random properties. These random properties are generated by a shift register having certain feedback as shown in Figure 4.12. Two signals from two different states of the shift register are MOD2 added to obtain a third signal, which is then fed back to another state of the shift register, thus accomplishing randomization. The number of feedback elements depends on the type of the function and length of the shift register. The total number of random sequences that can be generated by means of an m-bit shift register is given by

$$N = 2^m - 1 \tag{4.17}$$

For example, the 3-bit shift register shown in Figure 4.12 [2] generates $2^3 - 1 = 7$ random sequences. Similarly a 64-bit shift register generates $2^{64} - 1 = 1.84 \times 10^{19}$ random sequences. These random sequences repeat themselves with the same random pattern. Although numerous PN sequences are available, only a few of them are used for cellular communication because of their unique correlation properties. These unique codes are known as orthogonal codes and have zero cross-correlation

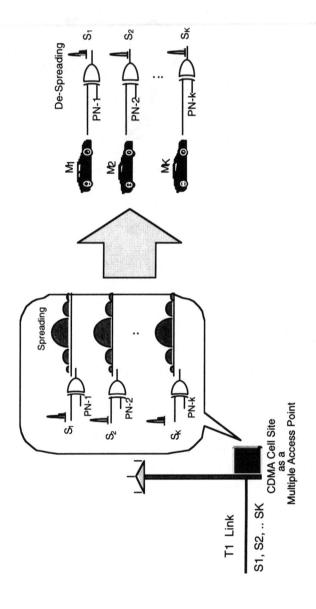

Figure 4.11 Conceptual representation of forward link DS-CDMA.

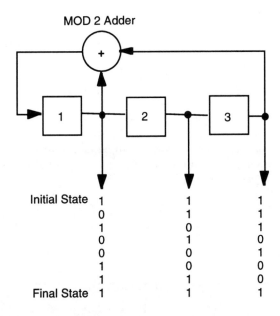

Figure 4.12 A 3-bit PN generator.

properties. The definition and generation of these codes along with their properties are briefly presented in the following subsections.

4.8.1 Orthogonal Codes

A pair of codes is said to be *orthogonal* if the cross-correlation is zero. For two *m*-bit codes: $x_1, x_2, \ldots, x_m$ and $y_1, y_2, \ldots, y_m$, this is given by

$$R_{xy}(0) = \sum_{i=1}^{m} x_i y_i = 0 \qquad (4.18)$$

For example, the cross-correlation between two 4-bit codes:

$$
\begin{aligned}
x &= \quad 0 \quad 0 \quad 1 \quad 1 \\
y &= \quad 0 \quad 1 \quad 1 \quad 0
\end{aligned}
$$

will be

$$
\begin{array}{cccc}
-1 & -1 & 1 & 1 \\
-1 & 1 & 1 & -1
\end{array}
$$

$$R_{xy}(0) = 1 - 1 + 1 - 1 = 0$$

where all 0s are replaced by -1. Similarly, the cross-correlation between a different pair of 4-bit codes: $x = 0\ 1\ 0\ 1$ and $y = 0\ 1\ 1\ 0$ is also zero as shown:

$$\begin{array}{rrrr} -1 & 1 & -1 & 1 \\ -1 & 1 & 1 & -1 \\ \hline \end{array}$$

$$R_{xy}(0) = 1 + 1 - 1 - 1 = 0$$

We also notice that an orthogonal code has an equal number of 1s and 0s. On the other hand, the following pair of codes is based on an equal number of 1s and 0s but their cross-correlation is not zero:

$$\begin{array}{rcccc} x = & 0 & 0 & 1 & 1 \\ y = & 1 & 1 & 0 & 0 \end{array}$$

will be

$$\begin{array}{rrrr} -1 & -1 & 1 & 1 \\ 1 & 1 & -1 & -1 \\ \hline \end{array}$$

$$R(xy) = -1 - 1 - 1 - 1 = -4$$

Therefore, this pair of codes is *not* orthogonal.

By continuing this process, we can show that an m-bit code has only m-orthogonal codes having an equal number of 1s and 0s per code. It follows that a 64-bit PN sequence provides only 64 orthogonal codes out of $2^{64} - 1 = 1.84 \times 10^{19}$ PN sequences. Thus we conclude that an orthogonal code has two basic properties:

1. An equal number of 1s and 0s;
2. Zero cross-correlation property.

A waveform that fulfills these requirements is a set of binary weighted waveforms, originally developed by Rademacher in 1922 [3] and shown in Figure 4.13(a). Two sets of n orthogonal codes are available of length $2n$ bits per code. Conversely, two sets of $n - 1$ orthogonal codes (dotted box) of length n are also available from this scheme.

4.8.2 Walsh Code

In 1923, J. L. Walsh introduced a complete set of orthogonal codes, based on rearranging the Rademacher code. These codes are also binary valued and obey the following two-step process:

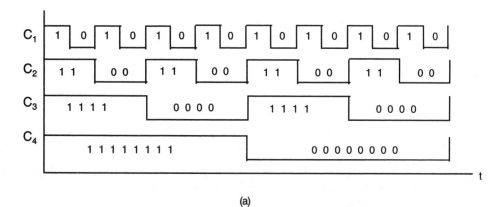

(a)

	Orthogonal Code (Set-1)								Orthogonal Code (Set-2)							
	1	2	3	4	5	6	7	8	9	10	11	12	13	14	15	16
C1	1	0	1	0	1	0	1	0	1	0	1	0	1	0	1	0
C2	1	1	0	0	1	1	0	0	1	1	0	0	1	1	0	0
C3	1	1	1	1	0	0	0	0	1	1	1	1	0	0	0	0
C4	1	1	1	1	1	1	1	1	0	0	0	0	0	0	0	0

(b)

Figure 4.13 (a) Binary weighted waveforms generating orthogonal codes. (b) Two sets of 8-bit orthogonal codes or two sets of 3-bit orthogonal codes (dotted box).

Step 1: Represent an $N \times N$ matrix as four quadrants:

N x N Matrix

1st Quadrant	2nd Quadrant
3rd Quadrant	4th Quadrant

Step 2: Make the first, second, and third quadrants identical and invert the fourth:

$$
\begin{array}{c|c}
b & b \\
\hline
b & \bar{b}
\end{array}
$$

where b is a bit that can be either 0 or 1. This process generates N, N-bit Walsh codes. For example, for a 2×2 matrix (two 2-bit Walsh codes), we have:

$$
\begin{array}{c|c}
b & b \\
\hline
b & \bar{b}
\end{array}
\quad = \quad
\begin{array}{c|c}
1 & 1 \\
\hline
1 & 0
\end{array}
\quad \text{or} \quad
\begin{array}{c|c}
0 & 0 \\
\hline
0 & 1
\end{array}
$$

i.e.   Code-1= 1 1   or 0 0
       Code-2 = 1 0   or 0 1

Similarly, for a 4×4 matrix (four 4-bit Walsh codes), we obtain:

$$
\begin{array}{cc|cc}
b & b & b & b \\
b & \bar{b} & b & \bar{b} \\
\hline
b & b & \bar{b} & \bar{b} \\
b & \bar{b} & \bar{b} & b
\end{array}
\quad = \quad
\begin{array}{cc|cc}
1 & 1 & 1 & 1 \\
1 & 0 & 1 & 0 \\
\hline
1 & 1 & 0 & 0 \\
1 & 0 & 0 & 1
\end{array}
\quad \text{or} \quad
\begin{array}{cc|cc}
0 & 0 & 0 & 0 \\
0 & 1 & 0 & 1 \\
\hline
0 & 0 & 1 & 1 \\
0 & 1 & 1 & 0
\end{array}
$$

where

Code-1 = 1 1 1 1 or 0 0 0 0

Code-2 = 1 0 1 0 or 0 1 0 1

Code-3 = 1 1 0 0 or 0 0 1 1

Code-4 = 1 0 0 1 or 0 1 1 0

This principle can be extended to generate an $N \times N$ matrix where N is binary valued.

4.9 NORTH AMERICAN DS-CDMA STANDARD

The North American DS-CDMA standard [1] is a dual-mode wideband spread-spectrum cellular system, in which one mode of operation is AMPS and the other mode of operation is CDMA. The CDMA standard is the subject of discussion in the following sections.

4.10 CDMA FREQUENCY BANDS

The existing 12.5-MHz cellular bands are used to derive 10 different CDMA bands, 1.25 MHz per band, as shown in Figure 4.14. Each of these bands supports 64 Walsh codes: W0, W1, . . . , W63 where each code is designated as a channel. These codes are not permitted to be reused in the same band, but they can be reused in another band. On the other hand, several frequency bands are permitted to be used in the same cell or a sector for capacity enhancement, as long as the frequencies are different.

4.11 CDMA CODES AND ITS USAGE

There are three different types of PN codes used in IS-95 CDMA:

1. Walsh code;
2. Long PN code;
3. Short PN code.

A brief description of these codes and their usage is presented next.

Figure 4.14 CDMA frequency bands.

4.11.1 Walsh Code

The Walsh code, also known as the Hadamard code, is a set of 64 orthogonal codes, stored in read-only memory (ROM). Their purpose is to provide:

1. Forward channel spreading over the 1.2288-Hz band;
2. Unique identification to a mobile.

The chip rate (code rate) of a Walsh code is 1.2288 megachips per second (Mcps).

The four different types of forward channels are designated as follows:

1. *Pilot channel:* W0 (Walsh code 0);
2. *Paging channel:* W1 to W7 (unused paging codes can be used for traffic);
3. *Sync channel:* W32;
4. *Traffic channel:* W8 to W31 and W33 to W63.

4.11.2 Long PN Code

The long PN code is generated from a 42-bit shift register having $2^{42} - 1 = 4.398 \times 10^{12}$ different codes. These codes are used for:

1. Base-band data scrambling in the forward path;
2. Base-band data spreading in the reverse path.

Note that the Walsh code is *not* used in the reverse path for spreading. However, a mobile acknowledges the assigned Walsh code by sending a 6-bit code that identifies one of 64 codes ($2^6 = 64$) that is assigned to it. The chip rate of the long PN code is 1.2288 Mcps.

4.11.3 Short PN Code

The short PN code is generated from a pair of 15-bit shift registers having a quadrature pair of $2^{15} - 1 = 32,767$ codes. These codes are used for cell identification in a reused cell. The chip rate of the short PN code is 1.2288 Mcps.

4.12 CDMA AIR-LINKS

CDMA air-link is based on a forward link and a reverse link, separated by 45 MHz as shown in Figure 4.15. The forward link is comprised of four different link

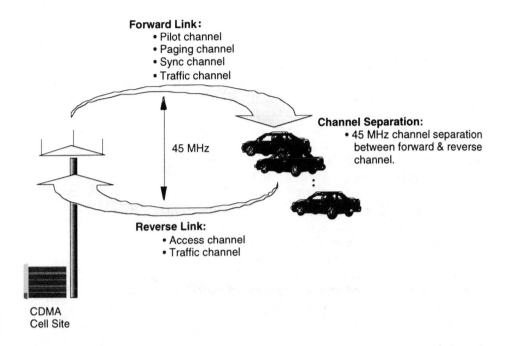

Figure 4.15 CDMA air-link structure.

protocols: (1) pilot channel, (2) paging channel, (3) sync channel, and (4) traffic channel. The reverse link is based on two different link protocols: (1) access channel and (2) traffic channel.

Each of these channels is spread by means of an appropriate Walsh code at a fixed chip rate of 1.2288 Mcps. The composite data are then MOD2 added with a quadrature pair of PN sequences at the same rate (there is no further spreading). The function of this PN sequence $(2^{15} - 1)$ is to provide a unique identification of the cell site so that a channel can be reused in another cell. This is similar to DVCC (digital voice color code) in TDMA, which is used as a cochannel identifier. The aggregate data are then modulated before transmission.

Briefly, the link protocol is as follows:

- Mobile acquires phase, timing, and signal strength via the pilot channel.
- Mobile synchronizes via the sync channel.
- Mobile gets system parameters via the paging channel.
- Mobile and base station communicate over the traffic channels during conversation.
- Mobile and base station communicate over access and paging channels during system acquisition and paging.

4.12.1 Forward Pilot Channel

The forward pilot channel (Figure 4.16) is an all-zero, uncoded spread-spectrum channel, continuously transmitted by each base station. Each base station uses one of $2^{15} - 1$ PN sequences, having a unique offset (0 through 511) to identify a forward CDMA pilot channel. There are 512 offset values available that can be reused elsewhere in the same system. Being all-zero base-band data, the pilot channel effectively modulates the quadrature PN code. A postmodulation filter attenuates the out-of-band signals by 41 dB before transmission.

A summary of the pilot channel follows:

- Is transmitted at all times;
- Uses Walsh code W0;
- Provides phase reference to the mobile;
- Provides timing reference to the mobile;
- Provides signal strength to the mobile for channel acquisition;
- Reused in every cell and sectors with PN offset;
- Has $\approx 20\%$ radiated power in the pilot.

4.12.2 Forward Sync Channel

The forward sync channel (Figure 4.17) is a one-half rate convolutionally encoded, interleaved spread-spectrum channel, which is used by mobiles for initial synchronization. The base-band data are spread by Wash code W32, which is then MOD2 added with the quadrature PN code and finally modulated. A postmodulation filter attenuates the out-of-band signals by 41 dB before transmission. The sync channel

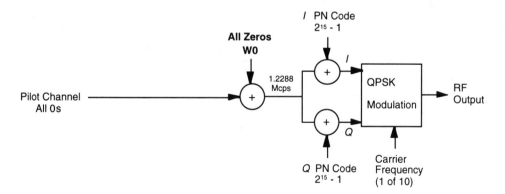

Figure 4.16 The pilot channel.

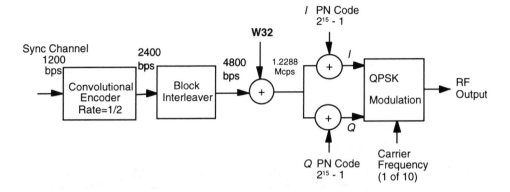

Figure 4.17 The sync channel.

uses the same PN offset as the pilot channel in a given cell. Having an identical PN offset, the synch channel aligns with the pilot channel immediately.

A summary of the sync channel follows:

- Bit rate = 1200 bps;
- Frame length = 26.666 ms;
- Walsh code = W32;
- Provides timing information to the mobile for synchronization;
- Provides pilot PN offset;
- Provides system time;
- Provides system and network IDs;
- Provides paging channel data rate.

4.12.3 Forward Paging Channel

The forward paging channel (Figure 4.18) is a one-half rate convolutionally encoded, interleaved spread-spectrum channel. The base station uses this channel to page mobiles and transmit system overhead messages. The base-band data are spread by one or more of Walsh codes W1–W7, which are then MOD2 added with the quadrature PN code and finally modulated. A postmodulation filter attenuates the out-of-band signals by 41 dB before transmission. The paging channel uses the same PN offset as the pilot channel in a given cell.

A summary of the paging channel follows:

- Bit rate = 9600 bps or 4800 bps [IS-95];
- Frame length = 20 ms;
- Walsh code = W1 to W7;

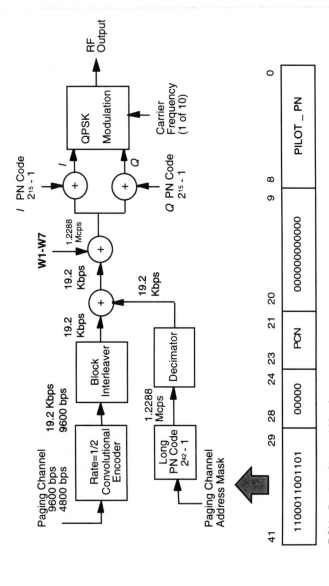

Figure 4.18 The paging channel.

- System parameter MSG;
- Neighbor list;
- Access parameter MSG;
- CDMA channel list;
- Overhead information;
- Pages a mobile;
- Assigns traffic channel to a mobile.

4.12.4 Forward Traffic Channel

The forward traffic channel (Figure 4.19) is a one-half rate convolutionally encoded, interleaved spread-spectrum channel. This channel is used for voice communication and signaling to a specific mobile during a call. The base-band data are spread by one of the traffic codes, MOD2 added with the quadrature PN code, and finally modulated. A postmodulation filter attenuates the out-of-band signals by 41 dB before transmission. The traffic channel uses the same PN offset as the pilot channel in a given cell.

A summary of the traffic channel follows:

- Bit rate = up to 9600 bps;
- Frame length = 20 ms;
- Used for voice communications and signaling;
- Uses Walsh codes W8–W31 and W33–W63.

4.12.5 Reverse Access Channel

The reverse access channel (Figure 4.20) is a one-third rate convolutionally encoded, interleaved spread-spectrum channel. It is used by a mobile to initiate a call with the base station and to respond to paging channel messages. Unlike the forward channel, the reverse channel is spread by one of $2^{42} - 1$ PN code (long PN code). The encoded spread-spectrum data are then MOD2 added with the quadrature PN code and finally modulated. A postmodulation filter attenuates the out-of-band signals by 41 dB before transmission.

A summary of the reverse access channel follows:

- Bit rate = 9600 and 4800 bps;
- System access;
- Traffic channel request;
- Call originations;
- Page response;
- System registration.

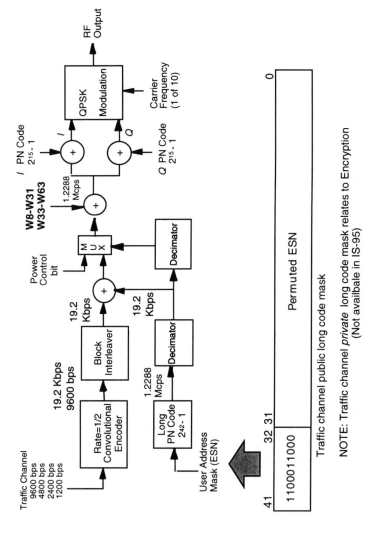

Figure 4.19 The forward traffic channel.

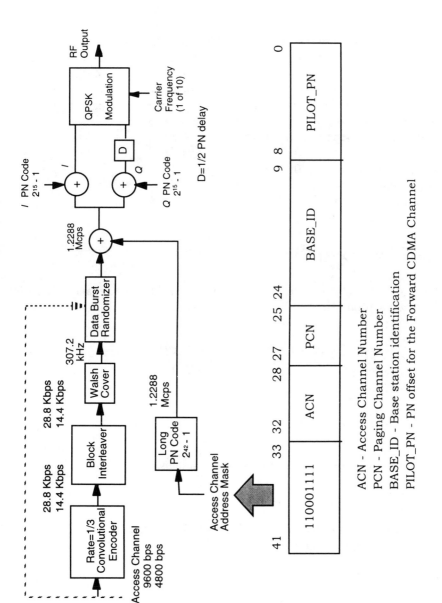

Figure 4.20 The reverse access channel.

4.12.6 Reverse Traffic Channel

The reverse traffic channel (Figure 4.21) is a one-third rate convolutionally encoded, interleaved spread-spectrum channel. It is used by a mobile for voice communication and signaling with the base. Unlike the forward traffic channel, the reverse traffic channel is spread by one of $2^{42} - 1$ PN code (long PN code). The encoded spread-spectrum data are then MOD2 added with the quadrature PN code and finally modulated. A postmodulation filter attenuates the out-of-band signals by 41 dB before transmission.

A summary of the reverse traffic channel follows:

- Bit rate = 9600, 4800, 2400, and 1200 bps;
- Voice communication;
- Responses to commands;
- Seeks information from the base.

4.13 CDMA SOFT CAPACITY

CDMA soft capacity (N_s) is a measure of maximum achievable capacity, given by the following equation [4]:

$$N_s = 1 + \frac{W/R_b}{E_b/N_o} \cdot (F/d) \cdot S \cdot H \tag{4.19}$$

where

$\quad N_s$ = number of simultaneous users per cell

$\quad W/R_b$ = process gain

$\quad E_b/N_o$ = ratio of energy per bit to noise spectral density

$\quad F$ = frequency reuse factor (<1)

$\quad d$ = voice duty cycle < 1

$\quad S$ = sectorization factor (value of 3 for tricellular plan)

$\quad H$ = soft hand-off factor.

With W = 1.25 MHz, R_b = 9600 bps, d = 0.45, F = 0.64, S = 3, and H = 1.5, we obtain the soft capacity as a function of E_b/N_o, shown in Figure 4.22. The total number of simultaneous users is maximized, depending on the minimum acceptable E_b/N_o. We find that two contradictory requirements exist for the capacity: (1) high capacity at the expense of E_b/N_o and (2) high E_b/N_o at the expense of capacity. It

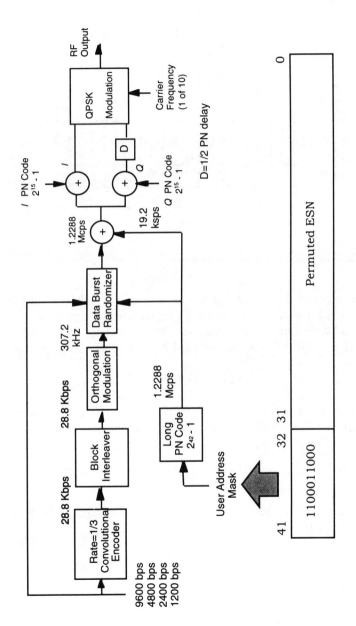

Figure 4.21 The reverse traffic channel.

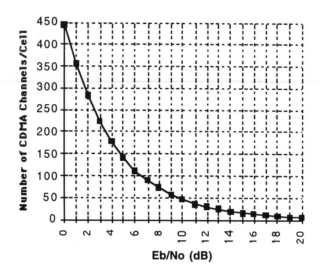

Figure 4.22 CDMA soft capacity as a function of E_b/N_o.

follows that a compromise is needed where the capacity is high and the E_b/N_o is the minimum for which the bit error rate (BER) performance acceptable.

Because all radios are based on some form of channel coding, we relate the coded bit error rate (BER$_c$) with the uncoded bit error rate (BER$_u$) as follows:

$$BER_c = m(BER_u)^n \tag{4.20}$$

where m and n are due to channel coding and the uncoded BER$_u$ is due to the modulation scheme. For QPSK, this may be approximated as

$$BER_u \approx \frac{\pi}{\sqrt{2}} e^{-E_b/N_o} \tag{4.21}$$

Combining (4.19), (4.20), and (4.21), we obtain the soft capacity as a function of coded bit error rate (BER$_c$):

$$N_s = 1 + nC \frac{W/R_b}{\ln\left[\dfrac{m(\pi/\sqrt{2})^n}{BER_c}\right]} \tag{4.22}$$

where $C = (F/d)$. Equation (4.22) shows that the CDMA capacity also depends on modulation scheme as well as on the channel coding scheme.

4.14 CDMA POWER CONTROL

4.14.1 Why Is Power Control Needed in CDMA?

CDMA is a multiple-access system in which several users (mobiles) have access to the same frequency band. As a result, the received signal strength will be different for different mobiles, resulting in near-far interference. Here, "near-far" refers to the ratio of signal strength from a near mobile to the signal strength from a mobile that is far away. This is critical for CDMA because the same frequency is shared by many mobiles. Near-far interference degrades performance, reduces capacity, and causes dropped calls. We examined this problem by means of Figure 4.23(a).

If the mobiles are permitted to transmit the same power from two different distances, the ratio of the received signals at the base station will be

$$\frac{\text{RSSI}_1}{\text{RSSI}_2} = \left(\frac{d_2}{d_1}\right)^{\gamma} \tag{4.23}$$

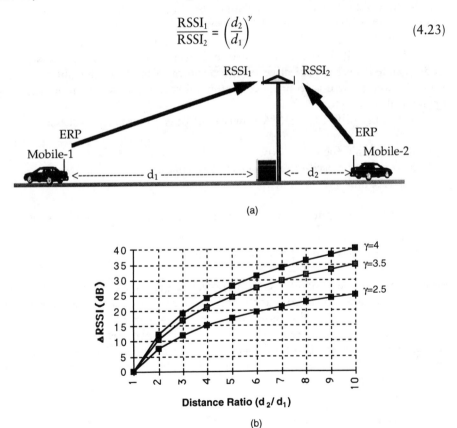

(a)

(b)

Figure 4.23 Illustration of near-far problem.

where

$RSSI_1$ = received signal from mobile 1

$RSSI_2$ = received signal from mobile 2

d_1 = distance between mobile 1 and the base station

d_2 = distance between mobile 2 and the base station

γ = path-loss slope (propagation environment).

Equation (4.23) implies that if $d_2 \neq d_1$ the received signal will be different for different mobiles depending on the propagation environment and the respective distances. This is illustrated in Figure 4.23(b) for several propagation environments. For example, if $d_2 = 4d_1$ and $\gamma = 4$ (typical dense urban environment), $RSSI_1$ from mobile 1 will be 256 times (24 dB) stronger than $RSSI_2$ from mobile 2, and the base station receiver will be unable to recover $RSSI_2$. Therefore, the transmitting power of each mobile has to be controlled so that its received power at the cell site is constant to a predetermined level, irrespective of the distance. Therefore, the objective of the mobile power control is to produce a nominal received power from all mobiles in a given cell or a sector.

According to IS-95, CDMA power control is a three-step process:

1. Reverse link open-loop power control (coarse);
2. Reverse link closed-loop power control (fine);
3. Forward link power control.

Each of these power control schemes is briefly described in the following subsections.

4.14.2 Reverse Link Open-Loop Power Control

Reverse link (mobile to base) open-loop power control is accomplished by adjusting the mobile transmit power so that the received signal at the base station is constant irrespective of the mobile distance; this is conceptually shown in Figure 4.24 where each mobile computes the relative path loss and compensates the loss by adjusting its transmitting power. The total received power at the cell site is the sum of all powers, which determines the system capacity. The dynamic range of this power control is 85 dB.

4.14.3 Reverse Link Closed-Loop Power Control

Reverse link closed-loop power control is accomplished by means of a power up or power down command originating from the cell site as shown in Figure 4.25. A

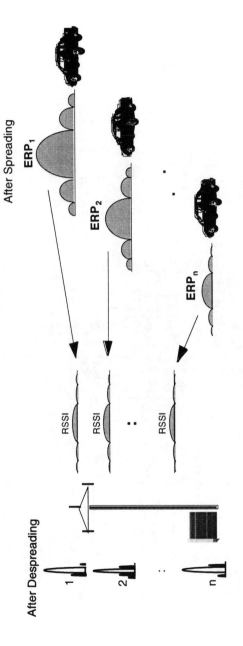

Figure 4.24 Reverse link open-loop power control scenario, showing each mobile controlling its own power so that the nominal received power at the cell site is content irrespective of the distance.

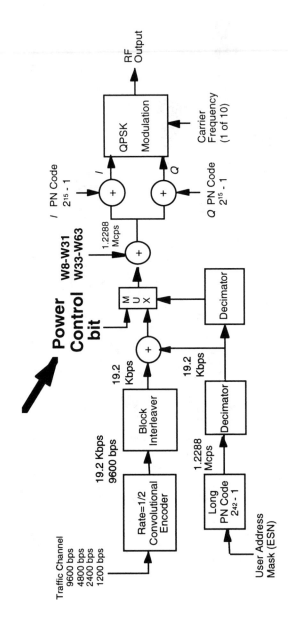

Figure 4.25 Reverse link closed-loop power control on the forward link. A single power control bit (1 for power down by 0.5 dB and 0 for power up by 0.5 dB) is inserted into the forward encoded data stream, every 1.25 ms.

single power control bit (1 for power down by 0.5 dB and 0 for power up by 0.5 dB) is inserted into the forward encoded data stream, every 1.25 ms. Upon receiving this command from the base station, the mobile responds by adjusting the power by an amount ±0.5 dB. The dynamic range of this power control is ±24 dB.

4.14.4 Forward Link Power Control

Forward link (base to mobile) power control is a one-step process. The base station controls its transmitting power so that a given mobile receives extra power to overcome fading, interference, BER, etc. In this mechanism, the cell site reduces its transmitting power while the mobile computes the frame error rate (FER). Once the mobile detects 1% FER, it sends a request to stop the power reduction. This adjustment process occurs once every 15 to 20 ms. The dynamic range of this type of power control is limited to only 6 dB in a 0.5-dB step because all mobiles are affected during this process. The methodology is conceptually presented in Figure 4.26.

4.15 CDMA CALL PROCESSING

4.15.1 Basic Call Processing Functions

CDMA call processing is based on the following link protocol:

- Pilot and sync channel processing;
- Paging channel processing;

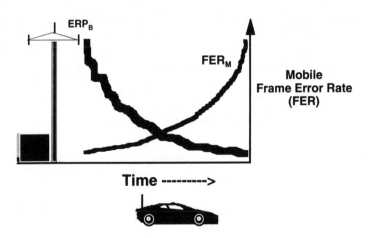

Figure 4.26 Forward link power adjustment. Base station reduces its transmit power until the mobile detects FER = 1%.

- Access channel processing;
- Traffic channel processing.

The associated call processing functions are executed by the base station and the mobile station in a coordinated manner. These steps are summarized in Table 4.1. The details can be obtained from IS-95 [1].

4.15.2 CDMA Hand-Off

In AMPS, hand-off is a process of changing the carrier frequency. Its primary purpose is to assign a new channel (frequency) when a mobile moves into an adjacent cell or a sector. Hand-off provides mobility and call continuity. This is accomplished by setting a hand-off threshold (i.e., if the received signal level is too low and reaches a predefined threshold, the system controller, namely, the mobile switching center, provides a stronger free channel over the voice channel). The voice is muted during this process by approximately 200 ms.

In CDMA, the purpose of hand-off is the same as in AMPS. However, there is no voice mutation during this process. This is an added advantage in CDMA, which is attributed to soft hand-off (make before break). This can be accomplished if the adjacent cell site uses the same frequency. Otherwise the hand-off is classified as hard hand-off. The three types of hand-off are described next.

CDMA to CDMA Soft Hand-Off

Soft hand-off is a process in which a mobile is directed to hand off to the same frequency, assigned to an adjacent cell or an adjacent sector without dropping the original RF link. The mobile keeps two RF links during the soft hand-off process. Once the new communication link is well established, the original link is dropped.

Table 4.1
Associated Call Processing Functions

Call Processing Channel	Base Station Functions	Mobile Station Functions
Pilot and synch channel	Continuously transmits pilot and sync	Acquires phase, timing, and power
Paging channel	Transmits overhead information, channel assignment	Acquires overhead and channel
Access channel	Base station monitors access channel	Responds to overhead, channel assignment and request access
Traffic channel	Voice communication	Voice communication

This process is also known as "make before break," which guarantees hand-off, provides diversity, enhances capacity, and reduces outage.

Mobile Activities

- Mobile station scans neighboring pilot channels and determines the strongest pilot.
- Mobile sends the information to the base via reverse traffic channel.

Base Activities

- Base station allocates a channel (Walsh code) over the forward traffic channel.
- Base station sends a hand-off direction message (cell/sector).

Mobile Activities

- Mobile decodes the new Walsh code.
- Mobile sends an acknowledgment.

CDMA to CDMA Hard Hand-Off

CDMA to CDMA hard hand-off is the process in which a mobile is directed to hand off to a different frequency assigned to an adjacent cell or a sector. The mobile drops the original link before establishing the new link. This process is also known as "break before make"; the voice is muted momentarily during this process.

CDMA to AMPS Hard Hand-Off

CDMA to AMPS hand-off is a process where a dual-mode mobile is directed to hand off to an AMPS channel. Voice is also muted momentarily during this process.

References

[1] IS-95, "Mobile Station–Base Station Compatibility Standard for Dual-Mode Wideband Spread-Spectrum Cellular Systems," TR-45, PN-3115, Electronic Industries Association Engineering Department, March 15, 1993.
[2] Smith, David R., *Digital Transmission Systems*, New York: Van Nostrand Reinhold Company, 1985.
[3] Elliot, Douglas F., and K. Ramamohan Rao, *Fast Transforms Algorithms, Analyses, Applications*, New York: Academic Press, 1982.
[4] Qualcomm, *The CDMA Network Engineering Handbook, Vol 1: Concepts in CDMA*, 1993.

CHAPTER 5
▼▼▼

PROPAGATION PREDICTION

5.1 INTRODUCTION

Radio-frequency (RF) propagation in a multipath environment is generally fuzzy because of irregular terrain, numerous RF barriers, and scattering phenomena. Building codes also vary from place to place and from one civil structure to another, requiring wide-ranging databases. As a result, a precise mathematical model is not available. Although sophisticated computer-aided prediction tools are available, these tools require user-defined clutter factors, thereby introducing impurities into the database. This results in an inevitable error in these prediction tools. Nevertheless, these tools are essential during the initial planning and deployment of cellular communication systems. Once the deployment is complete, various RF parameters such as RF coverage verification, interference reduction, and hand-off parameter adjustments are routinely carried out by means of a RF survey (drive test).

The purpose of this chapter is to provide a general understanding of RF propagation and the various attributes thereof. It begins with a brief overview of free-space propagation, followed by that of multipath propagation. Classical propagation models, such as Walfisch-Ikegami and Okumura-Hata, are then described. A simplified propagation model is then derived for low-power cellular and microcellular services. Finally, the mechanism of RF survey, both forward path survey and reverse path survey, is described with illustrations.

5.2 MECHANISM OF FREE-SPACE PROPAGATION

Electromagnetic waves differ in energy according to their wavelength (frequency). Their ability to propagate is also different in different propagation environments. In free space (vacuum) they are characterized by their ability to propagate without obstruction and without atmospheric effects. The path loss under these conditions is said to be *free-space loss*. For example, we consider an isotropic RF source that radiates electromagnetic energy uniformly in all directions as shown in Figure 5.1(a) in three-dimensional space. The radiating source is located at the center, which begins its emission at a given time. Maxwell's theory of electromagnetic radiation implies that the energy radiates uniformly in all directions, at the speed of light (3×10^8 m/s or 3.3 μs/km). This may be viewed as a concentric sphere, expanding in time and space.

Because it is difficult to represent time and space in four dimensions, we can represent this time-space relationship by means of a cross-sectional view of the energy sphere in two dimensions in space and one dimension in time as shown as a "cone" [1] in Figure 5.1(b), where time is represented by the vertical axis. There is no signal outside the cone since the velocity of the electromagnetic wave is constant. For example, if we assume that $d_1 = 1$ km, $d_2 = 2$ km, and $d_3 = 3$ km, the RF signal that originates at time $t = 0$, will arrive in those locations exactly after 3.3, 6.6, and 9.9 μs, respectively. This implies that the propagated signal exists within a space-time coordinate (d_i, t_i) where d_i is the location of the signal and t_i is the corresponding instant of time. The propagation delay is given by

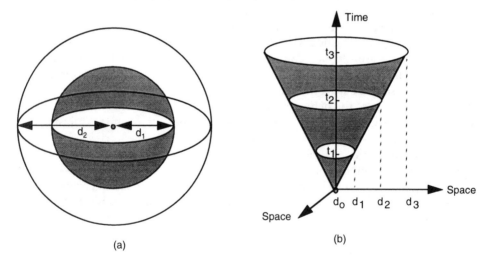

(a) (b)

Figure 5.1 (a) Electromagnetic radiation in three-dimensional space. (b) A cross-sectional view of radiation in time and space.

$$\tau_p \approx 3.3 \ \mu s/km \tag{5.1}$$

which is constant. This propagation delay is an important parameter in cellular communication systems. It determines the maximum cell size and intersymbol interference in digital cellular radios. For example, the signal that arrives at d_3, has a propagation delay of 9.9 μs. This delay determines the radius of the sphere, which has uniform signal strength throughout the surface of the sphere. In cellular communication this sphere is known as an ideal *cell*, and is shown in Figure 5.1(a). The total energy within the cell is constant irrespective of time and space.

The received power at the cell boundary may be obtained by assuming that the transmit antenna has a power of P_t, whose gain in a particular direction is G_t. Then the radiated power density at a given distance d will be given by

$$\rho = \frac{P_t G_t}{4 \pi d^2} \ W/m^2 \tag{5.2}$$

If a receive antenna is located at a distance d, whose gain is G_r and the effective area is A [2], where

$$A = G_r \frac{\lambda^2}{4 \pi} \tag{5.3}$$

then the received power P_r at the terminal of the receive antenna will be given by

$$P_r = \rho A = P_t G_t G_r \left(\frac{\lambda}{4 \pi d} \right)^2 \tag{5.4}$$

In the preceding analysis, we assumed that the transmission began at t_o and was received at t. The time difference $t - t_o$ is the propagation delay, which is

$$\tau_p = t - t_o \tag{5.5}$$

Thus, by knowing the start time, the propagation delay and hence the distance can be accurately determined.

Now, referring to (5.4) we find that the received signal attenuates as a square of the distance. Therefore, the free-space path loss formula (L_p), can be obtained as

$$L_p = 10 \log \left[\left(\frac{4 \pi d}{\lambda} \right)^2 \right] \tag{5.6}$$

or

$$L_p = 32.44 + 20 \log(f) + 20 \log(d) \tag{5.7}$$

where $\lambda = c/f$, $c = 3 \times 10^8$ m/s, the frequency (f) is measured in megahertz and the distance d is measured in kilometers. For an 800-MHz cellular band, we have 20 $\log(800) \approx 58$ dB and the free-space path loss becomes

$$L_p \text{ (dB)} = 90.88 + 20 \log(d) \tag{5.8}$$

or

$$L_p \text{ (dB)} = L_o + 10 \ \gamma \log(d) \tag{5.9}$$

which exhibits an equation of a straight line where L_o (= 90.88 dB) is the intercept and γ (= 2) is the slope. Later in this chapter we will see that all propagation models can be approximated as an equation of a straight line, which can be used to simplify the system design.

The preceding analysis assumes that the transmit power is the actual power that is transmitted from the tip of the antenna. For isotropic antenna this is defined as

$$P_t = \text{EIRP (effective isotropic radiated power) for isotropic antennas} \tag{5.10}$$

$$= \text{ERP (effective radiated power) for dipole antennas} \tag{5.11}$$

The received power P_r is defined as

$$P_r = \text{RSL (received signal level) for general purpose use} \tag{5.12}$$

$$= \text{RSSI (received signal level indicator) in IS-54 Standard [3]} \tag{5.13}$$

This notation is used throughout the rest of this book.

The relationship between EIRP and ERP is approximately as follows [4]:

$$\text{ERP} \approx \text{EIRP} + 2 \text{ dB} \tag{5.14}$$

This difference is due to the fact that an isotropic antenna radiates uniformly in all directions, whereas a vertically polarized dipole radiates more on the horizontal plane. Note that vertically polarized antennas are generally used for cellular communication systems.

Example

Given the following parameters:

$$\begin{aligned}
\text{Frequency} &= 850 \text{ MHz} \\
\text{EIRP} &= 45\text{W} \\
\text{Distance} &= 5 \text{ km} \\
\text{Radiation pattern} &= \text{Isotropic}
\end{aligned}$$

compute (a) the free-space path loss (L_p) and (b) the received power at 5 km.

Answer

(a) $L_p = 32.44 + 20 \log(850) + 20 \log(5)$
$\qquad = 32.44 + 58.59 + 13.98$
$\qquad = 105 \text{ dB}$
(b) RSL $= 45 \times 10^{-105/10}$
$\qquad = 1.423 \times 10^{-9} \text{ W}$
$\qquad = 10 \log(1.423 \times 10^{-9})$
$\qquad = -88.47 \text{ dBW} = -88.47 + 30 = -58.47 \text{ dBm}$
$\qquad$ or
$\qquad$ RSL $= 10 \log(45) - 105 \text{ dB} = 16.53 - 105 = -88.47 \text{ dBW} \ (-58.47 \text{ dBm})$

The preceding example illustrates the simplicity of using decibels. For instance, if we double the frequency, the path loss increases by $20 \log(2) = 6$ dB, whereas the received signal level reduces by 6 dB. Similarly, if we increase the distance by a factor of 2, path loss increases by 6 dB and the received signal level reduces by 6 dB.

5.3 MECHANISM OF MULTIPATH PROPAGATION

Multipath propagation is due to reflection, diffraction, and scattering of radiowaves caused by obstructions along the path of transmission. The magnitude of these effects depends on the type and total area of obstruction. For example, a plane surface of vast area will produce maximum reflection, whereas a sharp object such as a mountain peak or the edge of a building will produce scattering components with minimum effects known as the *knife-edge effect*. These spurious signals have longer path lengths than the direct signal. The associated magnitude and phase differences also vary according to the path length.

We begin our analysis using the familiar four-dimensional model shown in Figure 5.2(a). We assume that the radiating source is an RF pulse of duration Δt,

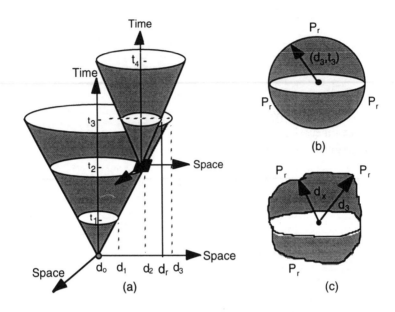

Figure 5.2 Mechanism of multipath propagation in time and space.

located at d_o, which begins its journey at time t_o^+ and the obstruction is located at d_2. Upon arrival at d_2, the signal encounters the obstruction, causing a secondary emission (Figure 5.2). The reflected component of the signal that originates at t_2 from location d_2 will arrive at a different location d_r at t_3, while the direct component of the signal arrives at d_3 at t_3 where $d_3 > d_r$. In other words, a receiver located at the line-of-sight (LOS) location d_3 will only detect the original signal if sampled exactly at t_3 with the following constraint of sampling duration (Δt):

$$\Delta t \leq 3.3(d_3 - d_r) \ \mu s \tag{5.15}$$

where d_3 is the distance between the transmitter and the receiver and d_r is the distance between the transmitter and an obstacle causing multipath radiation.

The corresponding cell at (d_3, t_3) is an uniform cell having a uniform received signal at the cell boundary, shown in Figure 5.2(b). This uniformity is due to the absence of multipath components at (d_3, t_3) where

$$(d_3, t_3) = \text{radius of the cell at the sampling instant } t_3 \tag{5.16}$$

On the other hand, if the signal is sampled at t_4 (see Figure 5.2(a)) with the following constraint of sampling duration (Δt),

$$\Delta t \geq 3.3(d_3 - d_r) \ \mu s \tag{5.17}$$

while the receiver is still at location d_3, the received signal will be due to both direct and multipath signals. The equal signal contour in this case will appear to be highly irregular due to destructive multipath components as in Figure 5.2(c). This is generally the case in mobile communication systems, causing delay spread [4,5].

5.3.1 Urban Propagation Environments

This is perhaps the most common and unpredictable propagation environment in cellular communication systems. Generally, it is known as an *urban environment* (Figure 5.3) where the density of civil structures varies from one urban environment to the other, requiring fine characterization of the urban environment such as dense urban, urban, suburban, etc. The received signals in these environments are a result of direct rays, reflected rays, and shadowing or any combinations of these signal components (neglecting atmospheric effects). Ground elevation also varies while the mobile is in motion. As a result the received signal varies erroneously due to interference. To accommodate these anomalies of propagation, sophisticated computer-aided prediction tools are available that are based on accurate terrain data and existing propagation models such as Okumura-Hata [6] and Walfisch-Ikegami [7]. These prediction models and their attributes are discussed later in this chapter.

5.3.2 Effects of Shadowing

Shadowing is a result of signal obstruction by trees, foliage, etc., as shown in Figure 5.4. Exact characterization of signal loss due to shadowing is difficult because the

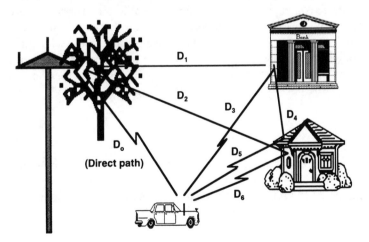

Figure 5.3 A typical urban environment showing shadowing and multipath.

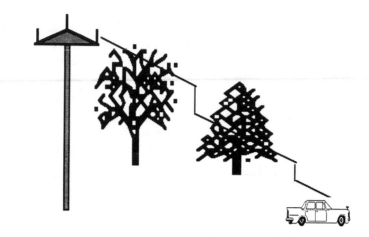

Figure 5.4 A mobile environment exhibiting the shadowing effect.

density of trees, foliage, etc., varies between place to place and season to season. The magnitude of this flat loss depends on the density of the forest as well as on seasonal variations, which should be obtained by means of measurements.

5.3.3 Propagation in Marine Environments

Sea ports, shipyards, dockyards, etc., are typical examples of cellular environments where the traffic is heavy (Figure 5.5). Multipath components in these environments are mainly due to ships, boats, and various metallic structures that are floating and nonstationary. Various civil structures along the shore are also responsible for generating multipath components. In the absence of multipath (see Figure 5.6), the received signal is mainly due to the direct path and reflections from water. Although path loss over water is similar to free-space path loss [4], these environments, especially sea ports, have to be properly characterized.

5.4 THE CELL

A cell is defined as a geographical area covered by RF signals. The shape and size of the cell depends on several parameters such as ERP, antenna radiation pattern, and propagation environments. Traditionally, a practical cell is assumed to be highly irregular having a regular received signal level (RSL) at the cell boundary as shown in Figure 5.7(a). On the other hand, the analytical cell, generally used for planning and engineering, is assumed to be a perfect hexagon, as shown in Figure 5.7(b). Evidently, there is a discrepancy between the analytical and the practical cell.

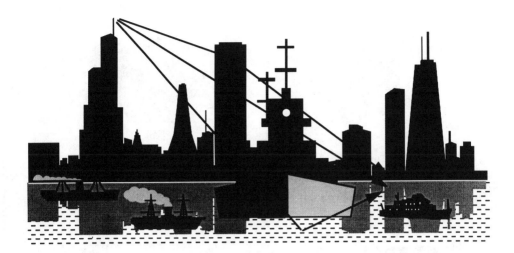

Figure 5.5 A marine environment exhibiting shadowing and multipath.

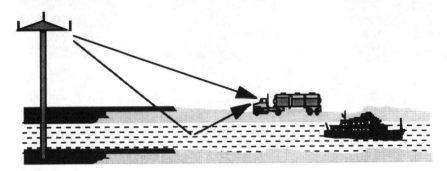

Figure 5.6 Propagation over water.

Conversely, one can define a cell as a perfect circle having different received signal levels at cell boundaries. This is illustrated in Figure 5.8(b) where the cell is always circular, whereas the received signal level at the cell boundary is irregular. A hexagonal cell having appropriate RSL values at cell boundaries can also be derived as shown in Figure 5.8(c). Because of this regularity, this approach enables quick and efficient radio deployment.

5.5 CLASSICAL PROPAGATION MODELS

Among numerous propagation models, the following are the most significant ones, providing the foundation of today's land-mobile communication services [6,7]:

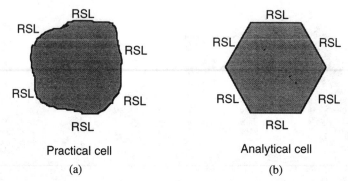

Practical cell

(a)

Analytical cell

(b)

Figure 5.7 (a) A practical cell having uniform RSL at cell boundaries. (b) A hexagonal cell, also having uniform RSL at cell boundaries.

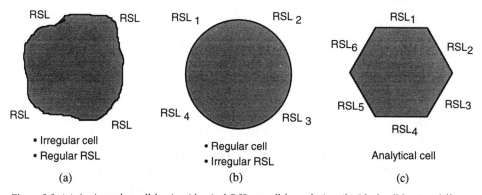

- Irregular cell
- Regular RSL

(a)

- Regular cell
- Irregular RSL

(b)

Analytical cell

(c)

Figure 5.8 (a) An irregular cell having identical RSL at cell boundaries. (b) Ideal cell having different RSL at cell boundaries. (c) Analytical hexagonal cell having appropriate RSL per side.

1. The Hata model;
2. The Walfisch-Ikegami model.

These prediction models are based on extensive experimental data and statistical analyses, which enable us to compute the received signal level in a given propagation medium. Many commercially available computer-aided prediction tools are also based on these models.

The usage and accuracy of these prediction models, however, depends chiefly on the propagation environment. For example, the standard Okumura-Hata model generally provides a good approximation in urban and suburban environments. On the other hand, the Walfisch-Ikegami model is applicable to dense urban environments. This model is also useful for microcellular services where antenna heights are generally lower than building heights, thus simulating an urban canyon environment.

The purpose of this section is to examine these models and represent them in a more convenient form for general understanding and application.

5.5.1 Hata Urban and Dense Urban Model

The Hata model [6] is based on experimental data collected from various urban environments having approximately 15% high-rise buildings. The general path-loss formula of the model is given by

$$L_p \text{ (dB)} = C_1 + C_2 \log(f) - 13.82 \log(h_b) - a(h_m) + [44.9 - 6.55 \log(h_b).] \log(d) + C_o \tag{5.18}$$

where

L_p = path loss, dB

f = frequency, MHz

d = distance between the base station and the mobile, km (1 km < d < 20 km)

h_b = effective height of the base station, m (30 m < h_b < 30 m)

$a(h_m)$ = {1.1 log(F) − 0.7}h_m − {1.56 log(F) − 0.8} for urban
= 3.2[log{11.75h_m}]^2 − 4.97 for dense urban

h_m = mobile antenna height (1 m < h_m < 10 m)

C_1 = 69.55 for 150 MHz ≤ f ≤ 1000 MHz
= 46.3 for 1500 MHz ≤ f ≤ 2000 MHz

C_2 = 26.16 for 150 MHz ≤ f ≤ 1000 MHz
= 33.9 for 1500 MHz ≤ f ≤ 2000 MHz

C_o = 0   for urban
= 3 dB for dense urban

Equation (5.18) may be expressed conveniently as

$$L_p \text{ (dB)} = L_o \text{ (dB)} + [44.9 - 6.55 \log(h_b).] \log(d) \tag{5.19}$$

or more conveniently as

$$L_p \text{ (dB)} = L_o \text{ (dB)} + 10\gamma \log(d) \tag{5.20}$$

Equation (5.20) exhibits an equation of a straight line where

$$L_o \text{ (dB)} = C_1 + C_2 \log(f) - 13.82 \log(h_b) - a(h_m) \tag{5.21}$$

is the intercept and γ is the path-loss slope, approximated as

$$\gamma = [44.9 - 6.55 \log(h_b)]/10 \qquad (5.22)$$

Equation (5.22) is plotted in Figure 5.9 as a function of base station antenna height. It shows that in a typical urban environment the attenuation slope varies between 3.5 and 4.

From (5.20) we also notice that the Okumura-Hata model also exhibits linear path-loss characteristics (equation of a straight line) as a function of distance where the attenuation slope is γ and the intercept is L_o. Because L_o is an arbitrary constant, we write

$$L_p \ (\text{dB}) \propto 10 \ \gamma \log(d) \qquad (5.23)$$

and in the linear scale,

$$L_p \propto \frac{1}{d^{\gamma}} \qquad \text{for } \gamma = 3.5 \text{ to } 4 \qquad (5.24)$$

5.5.2 Hata Suburban and Rural Model

Hata suburban and rural models are based on the urban model with the following corrections:

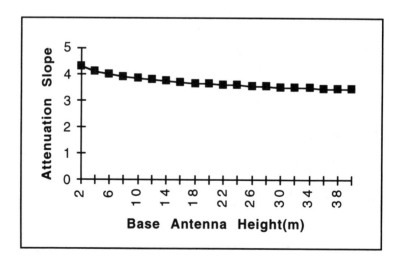

Figure 5.9 Attenuation slope as a function of base station antenna height in a typical urban environment (due to Hata).

$$L_p(\text{suburban}) = L_p(\text{urban}) - 2[\log(f/28)]^2 - 5.4$$
$$L_p(\text{rural}) = L_p(\text{urban}) - 4.78[\log(f)]^2 + 18.33 \log(f) - 40.94$$

These models also follow the equation of a straight line, thus affecting the intercept.

5.5.3 Walfisch-Ikegami Model

The Walfisch-Ikegami model [7] is useful for dense urban environments. This model is based on several urban parameters such as building density, average building height, and street widths. Antenna height is generally lower than the average building height, so that the signals are guided along the street, simulating an urban canyon type of environment.

For LOS propagation, the path-loss formula is given by:

$$L_p(\text{LOS}) = 42.6 + 20 \log(f) + 26 \log(d) \tag{5.25}$$

which can be described by means of the familiar "equation of a straight line" as

$$L_p(\text{LOS}) = L_o + 10 \ \gamma \log(d) \tag{5.26}$$

where L_o is the intercept and γ is the attenuation slope defined as

$$L_o = 42.6 + 20 \log(f)$$
$$\gamma = 2.6$$

Such a low attenuation slope in urban environments ($\gamma = 2.6$) is believed to be due to low antenna heights (below the rooftop), generating waveguide effects along the street. It follows that if a cell site is located at the intersection of a four-way street, the contour of constant path loss would look like a diamond as shown in Figure 5.10. Note that $\gamma = 2$ in free space.

For non–line-of-sight (NLOS) propagation, the path-loss formula is

$$L_p(\text{NLOS}) = 32.4 + 20 \log(f) + 20 \log(d) + L(\text{diff}) + L(\text{mult}) \tag{5.27}$$

where f is frequency, d is distance, $L(\text{diff})$ represents rooftop diffraction loss, and $L(\text{mult})$ represents multiple diffraction loss due to surrounding buildings. The rooftop diffraction loss is characterized as

$$L(\text{diff}) = -16.9 - 10 \log(\Delta W) + 10 \log(f) + 20 \log(\Delta h_m) + L(0) \tag{5.28}$$

where ΔW is the distance between the street mobile and the building, h_m is the mobile antenna height, Δh_m is $h_{\text{roof}} - h_m$ and $L(0)$ is the loss due to elevation angle.

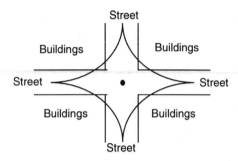

Figure 5.10 Diamond-shaped coverage in dense urban canyon.

Multiple diffraction and scattering components are characterized by following equation:

$$L(\text{mult}) = k_o + k_a + k_d \cdot \log(d) + k_f \cdot \log(f) - 9 \log(W) \qquad (5.29)$$

where

$k_o = -18 \log(1 + \Delta h_b)$
$k_a = 54 - 0.8(\Delta h_b)$ for $d \geq 0.5$ km
$\quad = 54 - 0.8(\Delta h_b)$ for $d \leq 0.5$ km
$k_d = 18 - 15(\Delta h_b/h_{\text{roof}})$
$k_f = -4 + 0.7[(f/925) - 1]$ for suburban
$\quad = -4 + 1.5[(f/925) - 1]$ for urban
W = street width
h_b = base station antenna height
h_{roof} = average height of surrounding small buildings ($h_{\text{roof}} < h_b$)
$\Delta h_b = h_b - h_{\text{roof}}$

We assume that the base station antenna height is lower than tall buildings but higher than small buildings.

Combining (5.27), (5.28), and (5.29), we obtain:

$$L_p(\text{NLOS}) = L_o + (20 + k_d) \log(d)$$
$$= L_o + 10 \ \gamma \log(d) \qquad (5.30)$$

The arbitrary constants are lumped together to obtain

$$L_o = 32.4 + (30 + k_f) \log(f) - 16.9 - 10 \log(w) + 20 \log(\Delta h_m) + L(0)$$
$$+ k_o + k_a - 9 \log(W)$$
$$\gamma = (20 + k_d)/10 \qquad (5.31)$$

Once again, the NLOS characteristics shown in (5.30) also exhibits a straight line with L_o as the intercept and γ as the slope.

The diffraction constant k_d depends on surrounding building heights, which vary from one urban environment to another, yielding a diffraction constant of a few meters to tens of meters. Typical attenuation slopes in these environments range from $\gamma = 2$ for $\Delta h_b/h_{\text{roof}} = 1.2$ to $\gamma = 3.8$ for $\Delta h_b/h_{\text{roof}} = 0$. This is shown in Figure 5.11.

5.5.4 Conclusions

From the preceding analysis we find that all the prediction models reduce to an equation of a straight line:

$$L_p \text{ (dB)} = L_o \text{ (dB)} + 10 \ \gamma \log(d) \tag{5.32}$$

with L_o as the intercept and γ as the slope.

Defining the received signal level as RSL (defined as RSSI in IS-54), and with 0-dB antenna gain, we can rewrite (5.32) as

$$RSL = ERP - L_p \tag{5.33}$$
$$= ERP - L_o - 10 \ \gamma \log(d)$$

which can be expressed in a more convenient form as

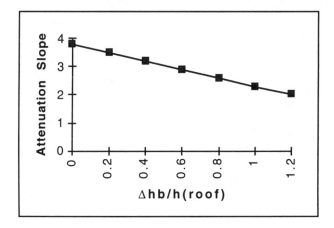

Figure 5.11 Attenuation slope as a function of base station antenna height in a typical dense urban environment (due to Walfisch-Ikegami).

$$d = 10^{(\text{ERP}-L_o-\text{RSL})/10\gamma} \qquad (5.34)$$

$$= 10^{-E/10\gamma}$$

where d is the distance, $E = \text{ERP} - L_o - \text{RSL}$, and γ is the path-loss slope, which is negative. Equation (5.34) indicates that there are three operating conditions:

1. $E = 0$ for which $d = 1$; independent of γ (multipath tolerance);
2. $E > 0$ for which $d < 1$; inversely proportional to γ (multipath attenuation);
3. $E < 0$ for which $d > 1$; proportional to γ (multipath gain or waveguide effect).

These parameters are further discussed in Chapter 8. The operating conditions are illustrated in Figure 5.12.

It follows that, if we can establish L_o and γ for different propagation environments (urban, suburban, rural) with reasonable accuracy, then the prediction mechanism and hence the deployment of cellular networks would be greatly enhanced.

5.6 APPROXIMATIONS FOR LOW-POWER MICROCELLS

5.6.1 Introduction

In many engineering investigations it is necessary to make predictions. Propagation prediction is no exception. In a multipath environment this signal suffers from unpredictable deep fades as described earlier. These anomalies of propagation brought our attention to the classical models developed by Walfisch-Ikegami and Okumura-Hata, described by a path-loss characteristic

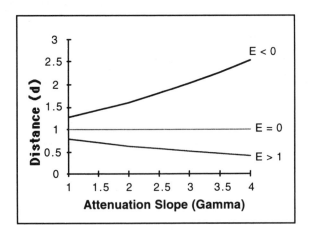

Figure 5.12 Relative coverage as a function of attenuation slope.

$$L_p \text{ (dB)} = L_o \text{ (dB)} + 10 \ \gamma \log(d)$$

or by a coverage equation

$$d = 10^{(ERP-L_o-RSL)/10\gamma}$$

where parameters are as described previously and γ is the attenuation slope, which is a function of the propagation medium. The preceding equation indicates that there is a unique set of design parameters for which the coverage is insensitive to γ. Because it is practically impossible to achieve such a goal, this section examines the *Fresnel zone breakpoint* [8,9] as a design parameter. A two-ray propagation model, based on this breakpoint, is presented for outdoor low-power microcells. A three-ray propagation model is then developed for indoor microcells. Empirical formulas are presented for cell radii and the associated link budget parameters. These prediction models generally apply to microcells because of increased LOS conditions.

5.6.2 Two-Ray Model

We begin our investigation by considering an outdoor propagation medium having a flat terrain as shown in Figure 5.13 where h_1 = transmit (Tx) antenna height, h_2 = receive (Rx) antenna height, d = antenna separation, d_1 = length of the reflected path, and d_2 = length of the direct path.

From plane geometry, the path differences between the direct and the reflected path can be estimated as

$$\Delta d = [(h_1 + h_2)^2 + d^2]^{1/2} - [(h_1 - h_2)^2 + d^2]^{1/2} \tag{5.35}$$

where $\Delta d = d_2 - d_1$. After some algebraic manipulation, (5.35) may be expressed as

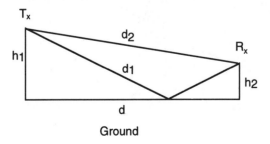

Figure 5.13 Two-ray model.

$$\Delta d = \frac{4h_1 h_2}{d[\{(h_1 + h_2)/d\}^2 + 1]^{1/2} + d[\{(h_1 - h_2)/d\}^2 + 1]^{1/2}} \tag{5.36}$$

where $(h_1 \pm h_2)/d \ll 1$ and the path difference in (5.36) reduces to

$$\Delta d \approx 2h_1 h_2/d \tag{5.37}$$

5.6.3 Effect of Fresnel Zone

In multipath environments, diffraction of radiowaves occurs when the wavefront encounters an obstacle. This can be examined by a model originally developed by Augustin-Jean Fresnel for optics. Fresnel postulated that the cross-section of an optical wavefront (electromagnetic wavefront) is divided into zones of concentric circles, separated by $\lambda/2$ (Figure 5.14) where λ is the wavelength.

The radius of the nth Fresnel zone is given by

$$R_n = [n\lambda(d_1 d_2/(d_1 + d_2)]^{1/2} \tag{5.38}$$

where

d_1 = distance between the transmitter and the obstruction
d_2 = distance between the receiver and the obstruction
$\lambda = c/f$
$n = 1$ for the first Fresnel zone
$n = 2$ for the second Fresnel zone.

From (5.38), we find that the Fresnel zone radius is inversely proportional to the square root of frequency. This implies that for a given antenna height, a high-

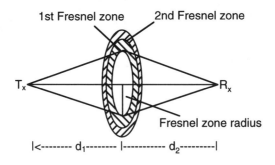

Figure 5.14 The Fresnel zone.

frequency signal will propagate further before the first Fresnel zone touches the ground (Figure 5.15(a)). Likewise, for a given frequency, a signal that radiates from a tall antenna will propagate further before the first Fresnel zone touches the ground (Figure 5.15(b)). In other words, diffraction of radiowaves depends on frequency as well as on antenna height.

When the path difference between the direct ray and the diffracted ray is $\lambda/2$, the diffraction is maximum. Thus from (5.37) we write:

$$\Delta d \approx 2h_1 h_2/d = \lambda/2 \tag{5.39}$$

resulting in

$$d_o = d = 4h_1 h_2/\lambda \tag{5.40}$$

This distance (d_o) is known as the *Fresnel zone breakpoint* [8,9], which is proportional to frequency and antenna height (Figure 5.16). The LOS path-loss slope within d_o is similar to free-space path loss since diffraction and multipath phenomena generally occur beyond this region.

5.6.4 An Alternate Proof

The analysis presented in the preceding section can be verified by combining the powers received from the direct path and the received path. Thus, referring to the two-ray model, the composite received power can be expressed as [4]

$$P_r = P_T(\lambda/4\pi d)^2[1 + e^{-j\Delta\theta}]^2$$
$$= [P_T(\lambda/4\pi d)^2] \cdot [4 \sin^2(\Delta\theta/2)] \tag{5.41}$$

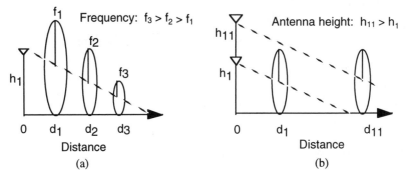

Figure 5.15 (a) A high-frequency signal propagates further before the first Fresnel zone touches the ground. (b) A signal radiating from a tall antenna propagates further before the first Fresnel zone touches the ground.

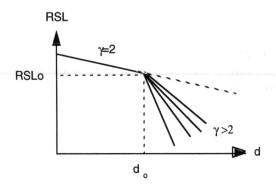

Figure 5.16 Illustration of Fresnel zone breakpoint.

where $\Delta\theta$ is the phase difference between the direct and the reflected path. In terms of path difference, it is given by

$$\Delta\theta = (2\pi/\lambda)\Delta d \tag{5.42}$$

Combining (5.41) and (5.42), we obtain:

$$P_r = [P_T(\lambda/4\pi d)^2] \cdot [4\sin^2(\pi/\lambda)\Delta d] \tag{5.43}$$

and with $\Delta d \approx 2h_1 h_2/d$,

$$P_r = [P_T(\lambda/4\pi d)^2] \cdot [4\sin^2\{(\pi/\lambda) \cdot 2h_1 h_2/d\}] \tag{5.44}$$

which is maximum for

$$(\pi/\lambda) \cdot 2h_1 h_2/d = (\pi/2) \tag{5.45}$$

or

$$d_o = d = 4h_1 h_2/\lambda \tag{5.46}$$

Under this condition, the received power can be obtained with the following constraints:

1. Antenna separation, $d \gg h_1$ and h_2;
2. Incident angle is negligible;
3. Phase difference $(\Delta\theta)$ is negligibly small.

Then (5.44) reduces to

$$P_r \approx [P_T(\lambda/4\pi d_o)^2] \tag{5.47}$$

which is the free-space loss. Note that d is now replaced by d_o, which is the Fresnel zone breakpoint. Therefore, path-loss characteristics within d_o will be similar to free-space path loss, that is, square law attenuation. The signal attenuates faster beyond d_o due to destructive multipath components, represented by (5.44) and plotted in Figure 5.17.

From the preceding analysis we conclude that there is a free-space path-loss region before the Fresnel zone breakpoint. After the breakpoint, the signal attenuates faster depending on the propagation medium. This breakpoint is a function of frequency and transmitting/receiving antenna heights.

Example

Outdoor LOS coverage given the following parameters:

$$\text{Frequency} = 900 \text{ and } 1900 \text{ MHz}$$
$$\text{Tx antenna height} = 30\text{m}$$
$$\text{Rx antenna height} = 1.5\text{m}$$

compute the breakpoint (d_o).

Answer

$$d_o = 4 \times 30 \times 1.5 \times 900 \times 10^6/3 \times 10^8$$
$$= 540 \text{ m @ 900 MHz}$$
$$d_o = 4 \times 30 \times 1.5 \times 1900 \times 10^6/3 \times 10^8$$
$$= 1140 \text{ m @ 1900 MHz}$$

These radii are suitable for PCS microcells.

5.7 IN-BUILDING RF COVERAGE

In-building coverage is a major concern among the service providers mainly because of high attenuation of signals through buildings and numerous scattering components within the building. As a result it is difficult to come up with a reliable RF prediction. In an effort to alleviate these problems, this section examines the Fresnel zone

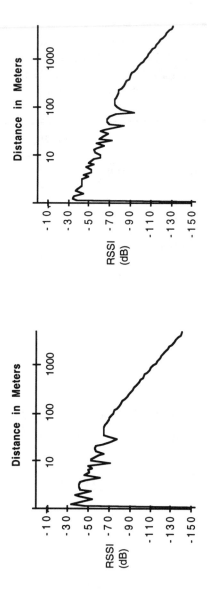

Figure 5.17 Received signal level as a function of distance showing the breakpoint (ERP = 1W).

breakpoint for two different deployment scenarios: (1) outdoor deployment, in which the base station antenna is located outside and (2) indoor deployment, in which the base station antenna is located inside the building. We show that, whereas a two-ray model applies to outdoor deployment, a three-ray model can be applied to in-building deployment where the location of the antenna with respect to floor and ceiling plays an important role in enhancing RF coverage and system performance. In the following, we examine outdoor deployment first, followed by indoor deployment.

5.7.1 Indoor Coverage From Outdoor Antennas (Two-Ray Approximations)

Civil structures create a complex environment in which RF signals suffer from high penetration losses due to outdoor multipath components, walls, metallic reinforcements, etc. The weak residual signal that finally survives suffers additional losses due to numerous scattering components within the building. As a result, the received signal within a building is generally weak. A possible measure to enhance the signal strength would be to avoid outdoor multipath components before building penetration. This can be accomplished by illuminating the building from within the Fresnel zone breakpoint. This is shown in Figure 5.18 where building 1 is within the breakpoint and building 2 is outside the breakpoint. Because the breakpoint is a function of the antenna height, several LOS buildings can be brought within this breakpoint simply by increasing the base station antenna height.

As an example, with F = 900 MHz, h_1 = 10m, and h_2 = 1.5m, we get $d_o = 4h_1h_2/\lambda$ = 180m. Only a few LOS buildings within 180m will be within this breakpoint. On the other hand, if h_1 = 30m, d_o becomes 540m; several LOS buildings can now be brought within the breakpoint.

5.7.2 Indoor Coverage From Indoor Antennas (Three-Ray Approximations)

We now consider an indoor propagation medium where the antenna is located inside the building, having ground reflections as well as reflections from the ceiling. This is shown in Figure 5.19 where

H = ceiling height
h_1 = transmitting antenna height
h_2 = receiving antenna height
d = antenna separation
D = direct path
d_1 = ground reflected path
d_2 = ceiling reflected path.

Because there are two dominant reflections, one from the floor and the other from the ceiling, our objective is to make these path differences identical so that the

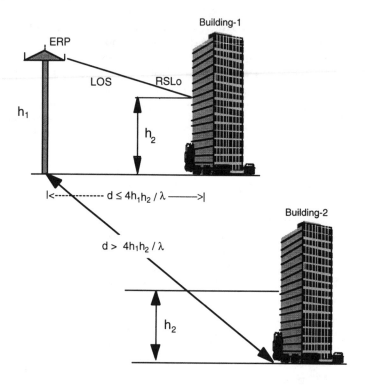

Figure 5.18 Indoor coverage from outdoor antenna. Building 1 is within the breakpoint and building 2 is outside the breakpoint, which can be brought within the breakpoint by raising the antenna h_1.

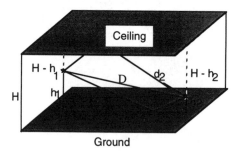

Figure 5.19 Three-ray indoor model.

first Fresnel zone breakpoint occurs at the same point. Note that reflections from side walls and their path differences are unpredictable since the portable cellular phone moves horizontally. Moreover, these horizontal components suffer from numerous losses due to penetration through glasses, porous materials, etc., before reflection. For these reasons, we ignore the horizontal components in the following analysis.

From plane geometry, the path differences between the direct and the vertical reflected paths (Figure 5.19) can be estimated as

$$\Delta d_1 = [(h_1 + h_2)^2 + d^2]^{1/2} - [(h_1 - h_2)^2 + d^2]^{1/2} \tag{5.48}$$

$$\Delta d_2 = [\{(H - h_1) + (H - h_2)\}^2 + d^2]^{1/2} - [\{(H - h_1) - (H - h_2)\}^2 + d^2]^{1/2} \tag{5.49}$$

where $\Delta d_1 = d_1 - D$ and $\Delta d_2 = d_2 - D$. After some algebraic manipulation, (5.48) and (5.49) can be expressed as

$$\Delta d_1 = \frac{4h_1 h_2}{d[\{(h_1 + h_2)/d\}^2 + 1]^{1/2} + d[\{(h_1 - h_2)/d\}^2 + 1]^{1/2}} \tag{5.50}$$

$$\Delta d_2 = \frac{4(H - h_1)(H - h_2)}{d[\{(2H - h_1 - h_2)/d\}^2 + 1]^{1/2} + d[\{(h_2 - h_1)/d\}^2 + 1]^{1/2}} \tag{5.51}$$

where $(h_1 \pm h_2)/d \ll 1$ and $2H - h_1 - h_2 \ll 1$. Thus we approximate as follows:

$$\Delta d_1 \approx 2h_1 h_2/d \tag{5.52}$$

$$\Delta d_2 \approx 2(H - h_1)(H - h_2)/d \tag{5.53}$$

Because it is desirable to have identical path differences, we write

$$\Delta d_1 = \Delta d_2 = \Delta d \tag{5.54}$$

for which we obtain the following identity:

$$H = h_1 + h_2 \tag{5.55}$$

The composite path difference then appears as

$$\Delta d = 2(H - h_2)h_2/d \tag{5.56}$$

which is a function of the ceiling height and the receiving antenna height. Within the first Fresnel zone, this path difference is exactly $\lambda/2$, for which there is maximum diffraction. Thus from (5.56) we write:

$$2(H - h_2)h_2/d = \lambda/2 \tag{5.57}$$

resulting in [10]:

$$d_o = d = 4(H - h_2)h_2/\lambda \quad \text{(indoor)} \tag{5.58}$$

Therefore, for optimum performance, the base station antenna should be located below the ceiling by h_2 where h_2 is the portable antenna height with respect to the floor. In other words, base station antenna height with respect to the ceiling should equal the portable antenna height with respect to the floor.

Example: Indoor LOS Coverage

Given the following parameters:

$$\text{Frequency} = 900 \text{ and } 1900 \text{ MHz}$$
$$\text{Ceiling height} = 4.5\text{m}$$
$$\text{Receiving antenna height} = 1.5\text{m}$$

compute the breakpoint (d_o).

Answer

Tx antenna height = $4.5 - 1.5 = 3\text{m}$ and

$$d_o = 4 \times 3 \times 1.5 \times 900 \times 10^6/3 \times 10^8$$
$$= 54 \text{ m @ 900 MHz}$$
$$d_o = 4 \times 3 \times 1.5 \times 1900 \times 10^6/3 \times 10^8$$
$$= 114 \text{ m @ 1900 MHz}$$

5.8 PREDICTION MODEL

We have established that the path-loss slope (γ) within the zone breakpoint is constant where $\gamma = 2$ before the breakpoint and $\gamma > 2$ after the breakpoint. Consequently, a cell whose radius is within this region remains insensitive to the propagation medium, since diffraction and other scattering phenomena occur beyond the Fresnel zone breakpoint.

From the equation of a straight line with RSL_o as the intercept at d_o, and γ as the slope beyond d_o (Figure 5.16), we obtain:

$$RSL = RSL_o - 10 \ \gamma \log(d/d_o) \qquad (5.59)$$

where

γ = attenuation slope
RSL = received signal level, dBm
d = range at the desired RSL, dBm
RSL_o = received signal level at the breakpoint d_o, dBm
d_o = Fresnel zone breakpoint.

Solving for d, we obtain the cell radius as a function of the breakpoint:

$$d = d_o 10^{(RSL_o - RSL)/10\gamma} \qquad (5.60)$$

which is similar to the one predicted by Walfisch-Ikegami and Okumura-Hata, except that we have introduced a variable d_o. Thus the exponent in (5.60) vanishes for RSL_o = RSL and the cell radius reduces to d_o, which is a function of frequency and transmitting/receiving antenna height. Within d_o, the cell is insensitive to the propagation environment where the RF signal attenuates gracefully with $\gamma = 2$. The path loss PL_o and the corresponding RSL_o within this region can now be obtained from the free-space path loss formula as

$$PL_o(\text{path loss}) = 32.44 + 20 \ \log(f) + 20 \ \log(d_o) \qquad (5.61)$$
$$RSL_o(d = d_o) = ERP - PL_o \qquad (5.62)$$

Combining (5.61) and (5.62), we obtain:

$$RSL_o = ERP - \{32.44 + 20 \ \log(f) + 20 \ \log(d_o)\} \qquad (5.63)$$

Substituting (5.63) into (5.60), we obtain the general prediction formula as:

$$d = d_o 10^{[ERP - \{32.44 + 20 \ \log(f) + 20 \ \log(d_o)\} - RSL]/10\gamma} \qquad (5.64)$$

An inspection of (5.64) reveals that there are three operating conditions that depend on the exponent: (1) The exponent is positive (+ve), for which d decreases with γ; this condition may be classified as *multipath attenuation*; (2) the exponent vanishes, for which d (= d_o) is insensitive to γ; this condition may be classified as *multipath tolerance*; (3) the exponent is negative (–ve), for which d increases with γ; this condition may be classified as *multipath gain*. These operating conditions are briefly described in the following sections.

5.8.1 Multipath Attenuation

Multipath attenuation is due to destructive interference after the breakpoint, where the reflected and diffracted components are >90 deg out of phase. Under this condition the link budget can be calculated by setting the exponent of (5.64) to +ve; that is,

$$\text{ERP} - [\{32.44 + 20 \log(f) + 20 \log(d_o)\} - \text{RSL}]/10\gamma > 0$$

for which, $d > d_o$ and the coverage becomes sensitive to γ. For a given propagation medium, this can be written as

$$\text{ERP} = 32.44 + 20 \log(f) + 20 \log(d_o) + 10 \ \gamma \log(d/d_o) + \text{RSL} \qquad (5.65)$$

Today's cellular communication systems fall largely into this category.

5.8.2 Multipath Tolerance

There is a unique combination of design parameters for which the exponent of (5.64) vanishes. The corresponding link budget becomes:

$$\text{ERP} = 32.44 + 20 \log(f) + 20 \log(d_o) + \text{RSL}_o \qquad (5.66)$$

and the coverage reduces to:

$$
\begin{array}{ll}
d = d_o = 4h_1 h_2 f/c & \text{(outdoor LOS coverage)} \\
d = d_o = 4(H - h_2)h_2 f/c & \text{(indoor LOS coverage)}
\end{array} \qquad (5.67)
$$

where $\text{RSL} = \text{RSL}_o$ and $H = h_1 + h_2$.

5.8.3 Multipath Gain

Multipath gain is due to constructive interference (waveguide effect), where the reflected and diffracted components are <90 deg out of phase and multiple reflections form a strong composite signal. Under this condition, the link budget can be calculated by setting the exponent of (5.64) to −ve:

$$\text{ERP} - [\{32.44 + 20 \log(f) + 20 \log(d_o)\} - \text{RSSI}]/10\gamma < 0 \qquad (5.68)$$

for which $d < d_o$. The path-loss slope under this condition is generally <2, which means that propagation is better than free space. Path-loss slope as low as 1.7 has been reported recently [11].

Example

Using the following set of indoor design parameters

Ceiling height H = 4.5m
h_2 = 1.5m (fixed)
h_1 = 4.5 − 1.5 = 3m (fixed)
F = 1900 MHz
RSL = −70 dBm

we obtain the set of curves shown in Figure 5.20(a). The ERP for which the cell is insensitive to propagation medium is 10mW. The corresponding cell radius is 114m.

For the same design parameters with twice the transmit antenna height (6m), the outdoor cell acquires twice the coverage (228m) (Figure 5.20(b)). The corresponding ERP = 32mW. We also notice that, both indoor and outdoor, the cell becomes gradually more sensitive to the propagation medium as the transmitting power increases. This is due to multipath fading beyond the breakpoint. The waveguide effect is also plotted in the same figure, showing coverage gain in multipath environments.

5.8.4 LOS Hand-Off Mechanism

Hand-off is a process that allows a cellular mobile to move from cell to cell without service interruption. Thus, referring to Figure 5.21, we find that a mobile crossing the cell boundary experiences a rapid attenuation of signals due to a large attenuation slope ($\gamma > 2$). As a result a prompt hand-off will take place. Once the hand-off is complete, a ping-pong effect is unlikely, since the difference in the RSL between the old and the new cell increases rapidly. This will also reduce hand-off requests, thus enhancing the capacity.

5.8.5 LOS Interference

Interference is a major problem in cellular communication systems because of frequent reuse of channels (cochannel interference). In the proposed model, this problem is greatly reduced due to properties inherent in the system.

To illustrate the concept, let us consider a pair of cells, separated by a repeat distance D, with R being the cell radius, as shown in Figure 5.22. Since the cochannel site is located beyond the breakpoint, its signal at the serving site will suffer multipath attenuation. Thus, assuming six cochannel sites, the carrier-to-interference C/I ratio prediction equation can be modified as,

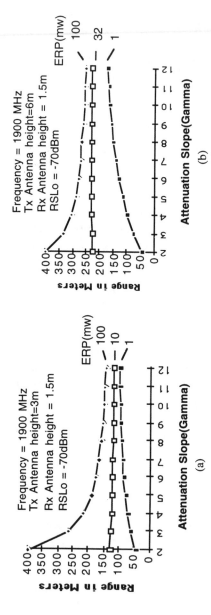

Figure 5.20 (a) Indoor cell radii as a function of propagation medium for $H = 4.5$m, $b_1 = 3$m, $b_2 = 1.5$m, and $F = 1900$ MHz. (b) Outdoor cell radii with twice the transmitting antenna.

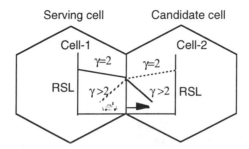

Figure 5.21 LOS hand-off scenario.

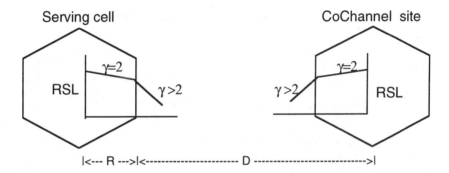

Figure 5.22 LOS cochannel interference.

$$\frac{C}{I} = 10 \log\left[\frac{1}{6}\left\{\frac{D^{(2+\Delta\gamma)}}{R^2}\right\}\right]$$

(5.69)

where $2 + \Delta\gamma$ is the path-loss slope beyond the breakpoint. As we can see, the C/I performance is enhanced if R is within the breakpoint and the reuse distance d is beyond the breakpoint. Stated differently, for a given C/I ratio, a channel can be reused more often, enhancing the system capacity. This analogy applies to those cells providing LOS conditions on major highways and streets.

5.8.6 Effects of Multipath and Shadowing

Let's assume that a mobile initiates a call from a LOS location and is moving toward a NLOS location. As long as the mobile is in line of sight and as long as $d \leq d_o$, the path-loss slope will be similar to free space (i.e., $\gamma \approx 2$). However, if there is an obstruction before the breakpoint, a sudden drop of signal strength will occur. This drop of signal strength is fairly steep and ranges from 15 to 25 dB depending on the multipath environment. Beyond this new breakpoint, the signal attenuates rapidly with $\gamma \gg 2$. This is conceptually shown in Figure 5.23 where

d_o = Fresnel zone breakpoint
Δd_o = distance between the obstruction and the Fresnel zone breakpoint
RSL_o = received signal level at the new breakpoint
ΔRSL = flat-loss due to shadowing.

From Figure 5.23, the received signal level can be determined as

$$RSL \approx (RSL_o - \Delta RSL) - 10 \; \gamma \log[d/(d_o - \Delta d_o)] \qquad (5.70)$$

where $\gamma > 2$ beyond the new breakpoint $d_o - \Delta d_o$. The corresponding distance can be computed as

$$d = (d_o - \Delta d_o)10^{[ERP-\{32.44+20 \log(f)+20 \log(d_o-\Delta d_o)\}-(RSL-\Delta RSL)]/10\gamma} \qquad (5.71)$$

This expression assumes that the antenna pattern is ideal. The unknown parameter γ determines the coverage. For γ insensitivity within the new breakpoint, the required link budget can be obtained by setting the exponent of (5.71) to zero, that is,

$$ERP = 32.44 + 20 \log(f) + 20 \log\{(d_o - \Delta d_o)\} + (RSL_o - \Delta RSL) \qquad (5.72)$$

where d_o is a function of antenna height and frequency and $RSL = RSL_o$ at d_o.

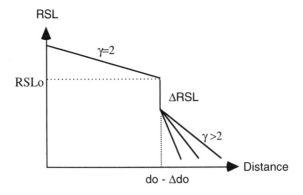

Figure 5.23 Loss due to free space, shadowing, and multipath.

5.8.7 Cell Site Location

To minimize multipath, it is desirable to maximize the Fresnel zone clearance. Maximization of Fresnel zone clearance also maximizes the breakpoint and hence the LOS coverage. Therefore, in a typical urban environment, the most desirable location would be the center of an intersection as shown in Figure 5.24.

For a given antenna height and transmitting frequency, the LOS received signal level at the cell boundary will be given by

$$RSL_o(LOS) = ERP - \{32.44 + 20 \log(f) + 20 \log(d_o)\} \tag{5.73}$$

The NLOS received signal may be written as

$$RSL(NLOS) = RSL_o(LOS) - \Delta RSL \tag{5.74}$$

where ΔRSL is the flat loss due to obstruction. Because it is desirable to have RSL $(NLOS) \geq RSL_{min}$, where RSL_{min} is the minimum specified level, the ERP is determined by RSL_{min}. The flat loss (ΔRSL) is an unknown parameter that should be established by means of experimental data.

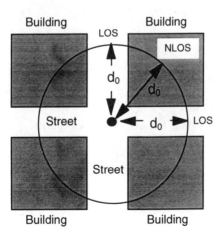

Figure 5.24 A cell in an urban environment showing LOS coverage along the streets and NLOS coverage due to obstruction.

5.8.8 Coverage Prediction

Using the following set of design parameters:

$$f = 1900 \text{ MHz}$$
$$\gamma = 2$$
$$h_1 = 8\text{m}$$
$$h_2 = 1.5\text{m}$$
$$\text{RSL (NLOS)} = -75 \text{ dBm}$$
$$\Delta\text{RSL} \approx 10 \text{ dB (say)}$$

we obtain:

$$d_o = 4h_1 h_2 f/c \text{ (outdoor)} = 304 \text{ m}$$

where d_o is in kilometers. Note that if the shadowing effect occurs before the breakpoint, d_o should modified as $d_o - \Delta d_o$ where Δd_o is obtained by measurement. The link budget can be obtained as

$$\text{ERP} = \text{RSL(NLOS)} + \Delta\text{RSL} + 32.44 + 20 \log(f) + 20 \log(d_o)$$
$$= 22.67 \text{ dBm (185 mW)}$$

With a 3 dB more flat loss ($\Delta\text{RSL} = 13$ dB), the ERP increases to 370mW. The ERP for an additional 3-dB increase of ΔRSL (= 16 dB) would still be $\leq$1W. Therefore, a 1W transmitter is a reasonable choice for PCS and microcellular services. The flat loss (ΔRSL) should be obtained by measurements. This is a small price to pay, since it is only necessary to obtain signal strengths from NLOS locations.

5.8.9 Receiver Sensitivity As a Function of Breakpoint

Receiver sensitivity is a measure of its ability to detect a weak signal. It is generally determined by the receiver thermal noise threshold $\{\eta(f)\}$. In decibels, this is given by the following equation:

$$\eta(f) = 10 \log\{(KTBW)NF\}$$
$$= 10 \log(KT) + 10 \log(BW) + NF(dB)$$
$$= -204 \text{ dB} + 10 \log(BW) + NF(dB) \tag{5.75}$$

where

K = Boltzmann constant $(1.3805 \times 10^{-23} \text{ J/K})$
T = absolute temperature (290 K)
BW = transmission bandwidth
NF (dB) = receiver noise figure.

For analog radio, it is determined by:

$$C/N = \text{RSL} - \eta(f)$$
where
$$\eta(f) = -204 + 10 \log(\text{BW}) + \text{NF} \qquad (5.76)$$
Therefore,
$$C/N = \text{RSL} + 204 - 10 \log(\text{BW}) - \text{NF}$$

where

$$\text{RSL} = \text{ERP} - \{\text{RSL}_o - 10 \ \gamma \log(d/d_o)\}.$$
$$C/N = \text{ERP} - \{\text{RSL}_o - 10 \ \gamma \log(d/d_o)\} + 204 - 10 \log(\text{BW}) - \text{NF} \qquad (5.77)$$

For digital radio, the performance is generally determined by E_b/N_o:

$$E_b/N_o = \frac{(\text{RSL})/r_b}{N_o} = \frac{\text{RSL}}{r_b N_o}$$

where

E_b = energy per bit
N_o = noise spectral density
r_b = bit rate.

In decibels, this may be expressed as

$$E_b/N_o = \text{RSL (dB)} - 10 \log(r_b) - 10 \log(N_o)$$
$$= \text{RSL (dB)} - 10 \log(r_b) + 204 \text{ dB} - \text{NF}$$

Substituting for RSL, we obtain:

$$E_b/N_o = \text{ERP} - \{\text{RSL}_o - 10 \ \gamma \log(d/d_o)\} - 10 \log(r_b) + 204 - \text{NF} \qquad (5.78)$$

5.9 LOS COVERAGE BEYOND THE BREAKPOINT

5.9.1 Introduction

The general formula for LOS coverage is given by

$$d = d_o 10^{[ERP-\{32.44+20\log(f)+20\log(d_o)\}-RSL]/10\gamma}$$

where γ is an unknown parameter except for free space. This is plotted in Figure 5.25 with the following design parameters:

$$
\begin{aligned}
\text{Frequency} &= 850 \text{ MHz} \\
\text{ERP} &= \text{variable} \\
\text{Tx antenna height} &= 30\text{m} \\
\text{Rx antenna height} &= 1.5\text{m} \\
\text{RSL} = \text{RSL}_o &= -73 \text{ dBm.}
\end{aligned}
$$

The breakpoint, for which the cell is insensitive to multipath, is given by $d_o = 510$m. The corresponding ERP is 17mW. We also notice that the cell radii gradually become sensitive to the propagation environment as the transmitting power increases. This is due to diffraction, scattering, and multipath fading beyond the breakpoint. Computation of LOS coverage now requires a precise knowledge of γ. Once the γ value is determined in a given propagation medium, the coverage can be easily computed.

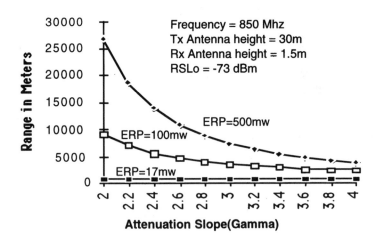

Figure 5.25 Cell radii as a function of propagation environment (γ).

Because it is practically impossible to characterize the entire spectrum of propagation environments, we consider a limited number of well-known propagation environments such as free space, rural, suburban, and urban and establish the respective attenuation slope (γ) by means of experimental data and regression analysis. For example, if the regression analysis yields $\gamma = 2.5$ for rural and $\gamma = 3$ for suburban, an experienced RF engineer could apply his/her best judgment (fuzzy logic) to establish a γ for an environment that falls between rural and suburban without additional measurements.

In the following, regression analysis is presented first, followed by the application of fuzzy logic. Regression analysis is used to characterize a limited number of known propagation environments and fuzzy logic is used to characterize an unknown environment from the set of known environments.

5.9.2 Regression Analysis

In experimental measurements we generally monitor two parameters simultaneously. One parameter is distance d, which is independent and user defined. The other parameter is RSL, which is random. We are interested in the dependence of RSL on d; we generally refer to it as the *regression curve of RSL on d*.

In the experiment we select n values of distances $d_1, d_2, \ldots, d_n$ and monitor the corresponding RSL value yielding a paired sample:

$$(d_1, \text{RSL}_1), (d_2, \text{RSL}_2), \ldots, (d_n, \text{RSL}_n) \tag{5.79}$$

In regression analysis the mean of RSL is a linear function of the distance, i.e.:

$$\overline{\text{RSL}} = \text{RSL}_o - \gamma d/d_o \tag{5.80}$$

In the logarithmic scale, it may be expressed as

$$\overline{\text{RSL}} \text{ (dB)} = \text{RSL}_o \text{ (dB)} - 10\ \gamma \log(d/d_o) \tag{5.81}$$

This is plotted in Figure 5.26 where RSL_o is the received signal level (intercept) at the breakpoint, d_o, and γ is the slope after d_o. The mean $\overline{\text{RSL}}$ is related to the random variable $(\text{RSL})_i$ as

$$(\Delta \text{RSL})_i = (\text{RSL})_i - \overline{\text{RSL}}$$
$$= (\text{RSL})_i - \text{RSL}_o + 10\ \gamma \log[(d_i/d_o)] \tag{5.82}$$

where $(\Delta \text{RSL})_i$ is the deviation of the random variable $(\text{RSL})_i$ from the mean $\overline{\text{RSL}}$. To fit a straight line, it is desirable to reduce $(\Delta \text{RSL})_i$. In other words, the sum of the square of this difference must be equal to zero. Thus we write:

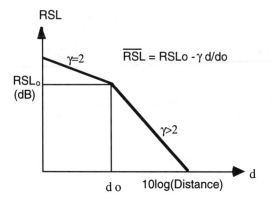

Figure 5.26 Regression curve in a given propagation environment.

$$\sum_{i=1}^{n} (\Delta RSL_i)^2 = 0 \qquad (5.83)$$

where

$$\sum_{i=1}^{n} (\Delta RSL_i)^2 = \sum_{i=1}^{n} [(RSL)_i - RSL_o + 10 \ \gamma \log(d_i/d_o)]^2 \qquad (5.84)$$

Then

$$\frac{\partial \Sigma (\Delta RSL_i)^2}{\partial RSL_o} = 0 \qquad \text{and} \qquad \frac{\partial \Sigma (\Delta RSL_i)^2}{\partial (d_i/d_o)} = 0 \qquad (5.85)$$

or

$$-2\sum_{i=1}^{n} [(RSL)_i - RSL_o + 10 \ \gamma \log(d_i/d_o)] = 0 \qquad (5.86)$$

$$-2\sum_{i=1}^{n} [(d_o/d_i)\{(RSL)_i - RSL_o + 10 \ \gamma \log(d_i/d_o)\}] = 0 \qquad (5.87)$$

which reduces to

$$\sum_{i=1}^{n} [(RSL)_i = n \cdot RSL_o - 10 \ \gamma \sum_{i=1}^{n} \log(d_i/d_o) \qquad (5.88)$$

$$\sum_{i=1}^{n} [d_o/d_i \cdot (RSL)_i = RSL_o \sum_{i=1}^{n} (d_o/d_i) - 10 \ \gamma \left[\sum_{i=1}^{n} (d_o/d_i) \cdot \log(d_i/d_o) \right] \qquad (5.89)$$

Solving (5.88) and (5.89), we obtain RSL_o and γ in a given propagation environment. The breakpoint is given by $d_o = 4h_1h_2/\lambda$, where h_1 and h_2 are the transmitting and receiving antenna heights, respectively. The above *method is known as the method of least squares* and can be used to fit a straight line to a given set of paired data (d_i, RSL_i).

5.9.3 Prediction of Random Data With Confidence

In statistical analysis we often induce a generalization from a set of random variables. For example, the regression analysis presented in the previous section is based on a set of paired data (d_i, RSL_i), where RSL_i is the random variable. It provides a point estimation from a random sample that enables us to estimate the received signal level at a given distance, with reasonable accuracy within a certain range of standard deviation. This range is known as the *confidence interval*. Once we have a confidence interval, we can reasonably assure ourselves that the mean of RSL will exist within this interval with a certain probability. This probability is known as the *confidence level*, which varies between 0 to 1 (0% to 100%).

Figure 5.27 shows the dependence of RSL as a function of distance, along with the regression line. Because we cannot expect to coincide the regression line with the actual straight line, we are instead interested in determining an interval within which the regression line may exist with a high probability.

Now we consider a set of random variables RSL_i having n sample values where $i = 1, 2, \ldots, n$. The distribution or the density of such a set of random numbers is generally approximated by a continuous curve known as a *normal distribution*. The equation that describes a normal distribution is given by [12,13]

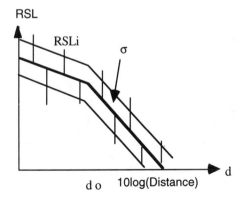

Figure 5.27 RSL as a function of distance. Solid lines are due to regression fit; do is the breakpoint and σ is the standard deviation.

$$f(\text{RSL}) = \frac{1}{\sigma\sqrt{2\pi}} \exp\left\{-0.5\left[\left(\frac{\text{RSL} - \overline{\text{RSL}}}{\sigma}\right)\right]\right\}^2 \tag{5.90}$$

where the *mean* is given by

$$\overline{\text{RSL}} = \frac{\text{RSL}_1 + \text{RSL}_2 + \ldots + \text{RSL}_n}{n} \tag{5.91}$$

and the *variance* is given by

$$\sigma^2 = \frac{(\text{RSL}_1 - \overline{\text{RSL}})^2 + (\text{RSL}_2 - \overline{\text{RSL}})^2 + \ldots + (\text{RSL}_n - \overline{\text{RSL}})^2}{n - 1} \tag{5.92}$$

where σ is the *standard deviation*.

The curve of Figure 5.28 is also known as the *Gaussian distribution* or a bell-shaped curve, which is symmetric with respect to the mean whose peak at $\overline{\text{RSL}} = 0$ increases as σ decreases.

Figure 5.29 shows the distribution curve for $\overline{\text{RSL}} \neq 0$. We notice that for a positive mean, the curve has the same shape but is shifted to the right; and for a negative mean, it is shifted to the left. This illustrates the fact that the *variance is the average dispersion from the mean*.

The probability density function of (5.90) is generally obtained from the standard table called *standard normal distribution*, or by means of a curve called the *cumulative distribution function* (Figure 5.30). Both are based on the following probability distribution function:

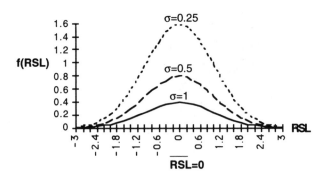

Figure 5.28 Normal distribution with zero mean (RSL = 0) and variable standard deviation (σ = 1, 0.5, and 0.25).

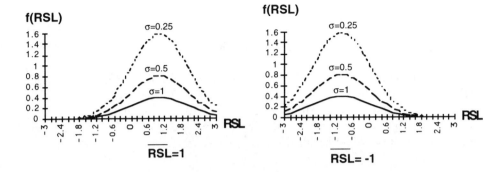

Figure 5.29 Normal distribution with RSL = 1 and RSL = −1 where σ is variable.

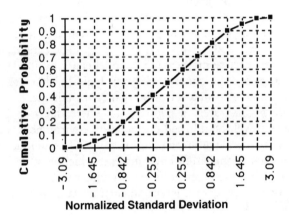

Figure 5.30 Cumulative probability distribution as a function of normalized standard deviation (z).

$$F(z) = \frac{1}{\sigma\sqrt{2\pi}}\int_{-\infty}^{z} \exp\left\{-0.5\left[\left(\frac{\text{RSL} - \overline{\text{RSL}}}{\sigma}\right)\right]\right\}^2 d(\text{RSL}) \qquad (5.93)$$

with $\overline{\text{RSL}} = 0$ and $\sigma = 1$. Then the random variable (RSL) can be estimated from the following normalized standard deviation z where σ is the measured standard deviation:

$$z = \left(\frac{\text{RSL} - \overline{\text{RSL}}}{\sigma}\right)$$

or

$$RSL = \sigma z = \overline{RSL} \qquad (5.94)$$

Thus if the desired received signal is = −80 dBm and the measured standard deviation $\sigma = 8$ dB, then for 80% probability ($z = 0.842$), the design criteria should be $(8 \times 0.842) - 80 = -73.26$ dBm. It means that 80% of the data will fall within the interval $-\sigma$ and $+\sigma$, i.e., within 8 dB. This interval is called the *confidence interval* and the probability (80%) is called the *confidence level*.

5.9.4 Application of Fuzzy Logic

Fuzzy logic is a branch of science that rationalizes uncertain events [14,15]. It manipulates vague concepts and provides a rational outcome. Fuzzy logic has been extensively used in many commercial products for temperature control, speed control, picture quality, etc. Today, application of fuzzy logic extends beyond commercial products where a precise mathematical model is not available. It is this "logic" that enables us to apply the concept of fuzzy logic to characterize an unknown propagation environment from a set of known environments.

This concept is illustrated in Figure 5.31 where the propagation medium is classified into several well-established propagation environments defined as a fuzzy set such as free space, rural, suburban, and urban [16]. A unique attenuation slope (γ) is assigned to each propagation environment, which is established by means of experiments as described in the previous section. Fuzzy logic reasons are then applied to determine γ for a propagation environment that is neither rural, suburban, or urban but closely resembles one of these environments.

Conceptually, fuzzy sets are continuous states having 50% typical overlap between adjacent sets. As a result, a given fuzzy set and its complementary value always sum to unity. For example, a pointer exactly on "suburban" means that the

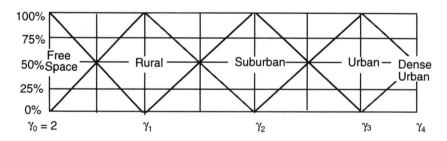

Figure 5.31 Classification of propagation environment. Attenuation slopes for rural, suburban, and urban environments are established by means of experiments.

attenuation slope is γ_2, whereas a pointer toward the left on the 50% mark would mean 50% rural and 50% suburban; the outcome then would be $\gamma_{1,2} = (\gamma_1 + \gamma_2)/2$. Similarly, a pointer toward the right on the 50% mark would mean 50% urban and 50% suburban; the outcome in this case would be $\gamma_{2,3} = (\gamma_2 + \gamma_3)/2$. This methodology is then used to determine an unknown propagation medium from a set of known propagation environments. Attenuation slopes for known environments (γ_1, γ_2, γ_3, etc.) are established my means of experimental data. The known environments may be classified as follows:

- *Free space:* A propagation environment having no obstructions.
- *Rural:* Open area, roads, and highways having no residential area.
- *Suburban:* Just outside of main downtown area, residential areas.
- *Urban:* Downtown area, having high-rise buildings on both sides of the streets.

The unknown environments may be obtained by means of the following *linguistic rules:*

- Rule 1: Free space–rural = 50% free space + 50% rural;
- Rule 2: Rural–suburban = 50% rural + 50% suburban;
- Rule 3: Suburban–urban = 50% suburban + 50% urban;
- Rule 4: Dense urban = denser than urban.

Apparently, these linguistic rules provide a fine-tuning of propagation environments that have already been established experimentally. There is nothing fuzzy about the logic itself; rather, it provides a better approximation.

5.10 COMPUTER-AIDED PREDICTION TECHNIQUES

5.10.1 Single Coverage Plot

With the advent of modern computers and software techniques, a large number of prediction tools are available today. These prediction tools generally begin with standard prediction models such as Okumura-Hata, Walfisch-Ikegami, or any user-defined propagation models, followed by additional information such as

- Terrain elevation data;
- Clutter factors (correction factors due to buildings, forests, water, etc.);
- Zone clearance;
- Antenna height, antenna pattern, ERP;
- Traffic distribution pattern, frequency planning, etc.

Terrain elevation data can be obtained from a topographical map if a manual calculation is desired. Conversely, it can be obtained in the form of a tape or floppy diskettes for computer-aided prediction. In either case, this information is a set of terrain data that generates a three-dimensional path profile as depicted in Figure 5.32.

The path profile is a graphic representation of the propagation path over an irregular terrain. This is an important parameter in determining the effective antenna height, which ensures adequate clearance from a given obstruction.

Various clutter factors, zone clearance, effective antenna height, ERP, etc., are then fed into the computer, which modifies the standard propagation model and computes the signal strength (RSL) at various (x,y,z) locations:

$$\mathrm{RSL}(x,y,z) \propto d^{-\gamma}(x,y,z) \tag{5.95}$$

where $\mathrm{RSL}(x,y,z)$ is the received signal level as a function of location, $d(x,y,z)$ is the distance with respect to location, and γ is the attenuation factor provided by the propagation model.

The outcome is a two-dimensional contour map in color graphics, which is conceptually shown in Figure 5.33. Signal strengths for various contours are also tabulated for quick reference. The distance from the cell center to any location in the coverage area is readily obtained from the display. This contour map is essentially the coverage of a cell for a given ERP in a given propagation environment, which indicates the received signal strength at various locations. Consequently, the shape and size of the cell will vary according to the propagation environment, clutter factors, and specific design parameters such as antenna height, antenna pattern, and ERP.

The preceding process is site specific, fairly accurate, and simple to use. For these reasons it is becoming increasingly popular among the cellular industries for

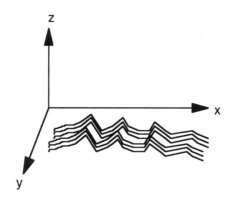

Figure 5.32 A three-dimensional path profile, where z is the elevation.

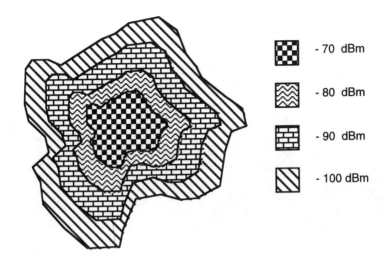

▨	- 70 dBm
▨	- 80 dBm
▨	- 90 dBm
▨	- 100 dBm

Figure 5.33 Computer-generated contour maps.

system planning and initial deployment. It also serves as an important tool during the quotation process.

5.10.2 Composite Coverage Plot

The composite coverage plot (map) is a collection of several contour maps plotted in a single sheet of paper. Today, most prediction tools are capable of plotting hundreds of contour maps in a single run where each contour represents a cell site. This is shown in Figure 5.34 for a cluster of seven cells. Note that a seven-cell cluster is a widely used cell plan among the cellular industries. It is also known as the $N = 7$ frequency plan, which we discuss in the next chapter.

Referring to Figure 5.34, we notice that there is wide variation of signal strength throughout the service area. We also notice a weak spot where the signal level is less than −100 dBm, which is generally not reliable, requiring an additional cell or a cellular repeater. Thus a composite coverage plot provides complete information about system health, which is essential for performance optimization.

5.10.3 Interference Analysis

Interference is a major concern in cellular communication systems. It is generally determined by the carrier-to-interference ratio (C/I), which in turn depends on frequency planning and antenna engineering. Adjacent channel interference (ACI) is also a concern if deployed in the adjacent site.

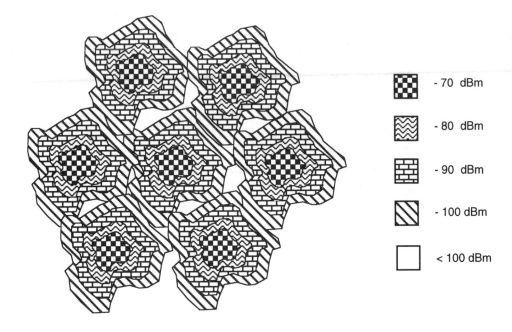

- 70 dBm

- 80 dBm

- 90 dBm

- 100 dBm

< 100 dBm

Figure 5.34 A composite coverage plot for a cluster of seven cells.

A cochannel interferer has the same nominal frequency as the desired frequency. It arises from multiple use of the same frequency. Thus, referring to the composite plot of Figure 5.35, we find that cochannel sites are located in the second cluster. For OMNI sites the theoretical cochannel interference is given by [4,5]

$$C/I = 10 \log \left[\frac{1}{j} \left(\frac{D}{R} \right)^{\gamma} \right] \qquad (5.96)$$

where

j = number of cochannel interferers (j = 1, 2, . . . , 6)
γ = propagation constant
D = frequency reuse distance
R = radius of the cell.

In a fully developed system, there are six interferers in the N = 7 frequency plan. Thus, with j = 6, γ = 4 (urban environment), and D/R = 4.58 [4,5], we obtain:

$$C/I = 18.6 \text{ dB} \qquad (5.97)$$

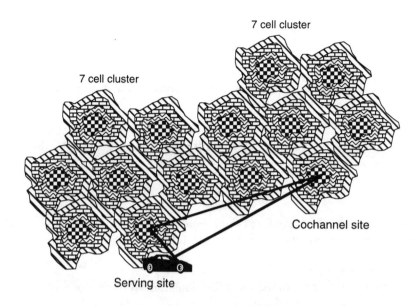

Figure 5.35 Cochannel interference evaluation scheme.

In computer-aided prediction technique, this value is readily available; it is site specific and more realistic due to terrain elevation data and clutter factors. Moreover, the location of potential problems due to cochannel interference is also identified in color graphics.

   Adjacent channel interference arises from energy slipover between two adjacent channels. This can be theoretically evaluated with the aid of Figure 5.36 where it is assumed that the ratio d_i/d_c varies as the mobile moves toward or away from the cell. Moreover, the out-of-band signals are attenuated by the postmodulation filter

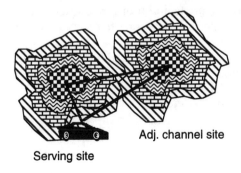

Figure 5.36 Adjacent channel interference evaluation scheme.

at least by 26 dB (EIA standard). Then the adjacent channel interference is computed as:

$$ACI = -10 \log\left[\frac{1}{j}\left(\frac{D}{R}\right)^{\gamma}\right] + \Delta dB \text{ (adjacent channel isolation)} \qquad (5.98)$$

where Δ dB is the adjacent channel isolation provided by postmodulation filters, which ≥ 26 dB. If the mobile is in the hand-off region, as in Figure 5.36, the distance ratio $D/R = 1$ and the worst case ACI becomes

$$ACI \text{ (worst case)} \approx -26 \text{ dB} \qquad (5.99)$$

These values are also computed and interference displayed by computer models.

In summary, the attributes of computer-aided prediction tools are:

- Produces composite coverage map;
- Generates interference plots, identifies potential problem areas;
- Identifies hand-off boundaries;
- Allows user to manually iterate design parameters such as frequency plan, antenna height, antenna downtilt, ERP, etc., to reduce interference and enhance performance;
- Enhances quotation and initial deployment processes.

5.11 RF SURVEY

An RF survey provides an effective measure of the propagated signal strength over an area of interest. It was a tedious task in the early days of cellular technology, owing to lack of automated measuring systems and software prediction tools. Today, RF survey is a mature technology in the cellular environment, in which sophisticated GPS receivers and software prediction tools are available for position location and RF prediction. It is an important tool for site selection, coverage verification, and cell site optimization.

Because RF propagation is characterized by antenna height, antenna radiation pattern, LOS propagation, NLOS propagation, shadowing, and multipath, careful planning and systematic data collection are essential for statistical analysis and proper interpretation of the results.

5.11.1 Antenna Calibration

The antenna radiation pattern plays an important role in determining the shape and size of the cell. Typical OMNI and directional patterns are shown in Figure 5.37.

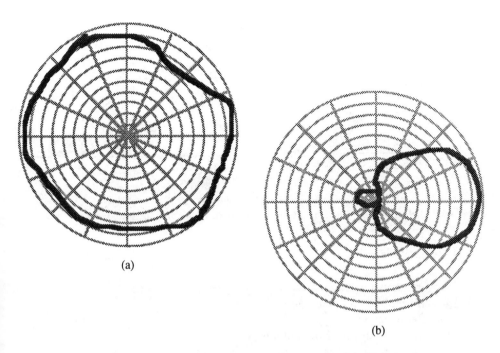

Figure 5.37 Typical OMNI and directional patterns. (a) OMNI pattern. (b) Directional pattern.

These patterns are generally available from the manufacturer. However, the pattern may also be verified without a laboratory environment, utilizing the benefit of the Fresnel zone breakpoint, as discussed briefly here.

The LOS path-loss characteristic within the zone breakpoint is linear and similar to free-space loss irrespective of the propagation medium. This important phenomenon enables us to verify the antenna radiation pattern in an outdoor environment, as long as the receiver has direct LOS and remains within the zone breakpoint. This mechanism is illustrated in Figure 5.38 where h_1 is the transmitting antenna height in meters, h_2 is the mobile antenna height in meters (typically 1.5m), and ERP is the effective radiated power.

The mobile is assumed to be in direct line of sight during the drive test up to the zone breakpoint, which is given by

$$d_o = \frac{4h_1 h_2}{\lambda} \tag{5.100}$$

A step-by-step characterization method follows:

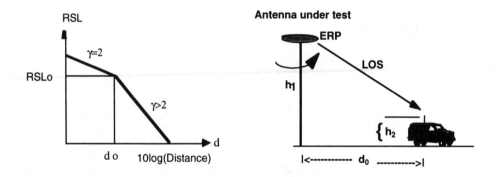

Figure 5.38 Method of measuring antenna radiation pattern within the zone breakpoint.

Step 1: Initial Preparations

- Measure transmit antenna height (h_1).
- Measure mobile antenna height (h_2).
- Note frequency (f).
- Note the orientation of the antenna.
- Find a LOS path that is valid up to d_o.

Step 2: Calculations

- Compute the breakpoint as

$$d_o = 4h_1h_2/\lambda$$

- Compute ERP as

$$ERP = 32.44 + 20 \log(f) + 20 \log(d_o) + RSL_o$$

where RSL_o is a desired threshold.

Step 3: Data Collection

- Set ERP as calculated in step 2.
- Measure and record RSL for every $\lambda/2$ up to do and obtain the regression line.
- Rotate antenna by, say, 45 deg and trace the same LOS route.
- Repeat the same process for a complete 360-deg rotation of the antenna.
- Draw the equal signal strength contour for the antenna radiation pattern.

The regression line for each trace for a given antenna orientation is given by

$$RSL = RSL_o + 10 \; \gamma \log(4h_1h_2/\lambda) \qquad (5.101)$$

where γ is the slope and RSL_o is the intercept. Within the breakpoint, $\gamma = 2$; after the breakpoint, $\gamma > 2$.

5.11.2 Forward Path RF Survey

Today's survey mechanism is based on an all-terrain vehicle that is equipped with field strength measuring equipments as shown in Figure 5.39. The vehicle is specially modified to absorb shock and to attenuate spurious RF signals originating elsewhere. The instantaneous received signal is recorded as a function of distance where the sampling interval is accurately recorded as the wheel turns. This is accomplished by

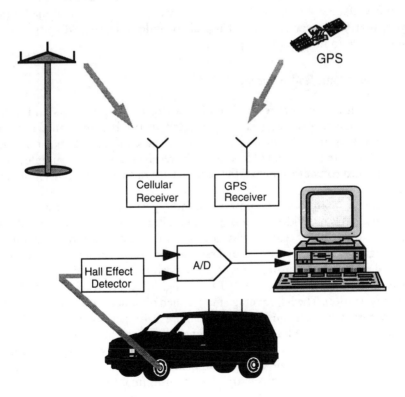

Figure 5.39 Forward path RF survey. The Hall effect detector controls the sampling rate and ensures data collection while the mobile is in motion.

means of a magnetic device connected to the wheel or to the speedometer. This arrangement controls the sampling rate and ensures data collection while the vehicle is in motion.

The global positioning system (GPS) receiver continuously monitors the position (x,y,z) of the mobile and the cellular radio provides the corresponding RSL. Both position (x,y,z) and RSL values are recorded into the computer. A coverage plot is then generated from a set of paired data (RSL_i, d_i).

Because there are three variables (x,y,z) in each coordinate, the GPS receiver tracks three satellites and simultaneously solves three equations to resolve each coordinate. Typical location accuracy, based on this scheme, is about 100m.

To improve the accuracy, a method known as differential GPS can be used. In this method a reference stationary GPS receiver, at a precise location on the ground, continuously monitors the received signals from all the geostationary satellites and computes the timing errors. It then passes this information to all the moving GPS receivers to make a correction. This method is accurate to less than 2m. Therefore, the accuracy of a RF survey can be increased by means of the second method. Note, however, that GPS receivers are not useful for indoor survey. Moreover, they are also susceptible to multipath components.

5.11.3 Reverse Path Radio Survey

The reverse path radio survey has been known of for several years and useful for its accuracy, due to diversity receivers located in the base station. However, its potential use was not fully exploited due to the lack of accurate position location systems. Now, with the advent of low-cost GPS receivers and MTX (mobile telephone exchange) based software support, the reverse path radio survey seems more feasible than ever before.

In general, both forward and reverse radio paths are identical with respect to propagation losses, provided there is no fading. But because fading is inevitable, the reverse path radio survey can provide a more effective and accurate measure of the received signal quality because of diversity at the base station. Thus we present a GPS-based survey mechanism [17] where the position (x,y,z) and the corresponding RSL values are transmitted to the MTX through diversity receivers, located in the cellular base station. The RF coverage map is then obtained from a set of RSL values.

This survey mechanism is shown in Figure 5.40. It is based on a GPS receiver for position location, a cellular radio for transmission of RSL, and MTX for coverage prediction. In this scheme, the GPS receiver continuously monitors the position (x,y,z) of the mobile and the cellular radio provides the corresponding RSL. Both position (x,y,z) and RSL values are then transmitted to the MTX via diversity receivers, located in the cellular base station. This is given by

$$RSL(x,y,z) \propto d^{-\gamma}(x,y,z) \qquad (5.102)$$

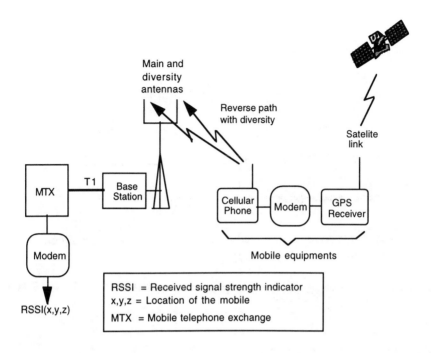

Figure 5.40 Reverse path radio survey.

where

RSL(x,y,z) = received signal strength as a function of location
$d(x,y,z)$ = distance
γ = attenuation factor (γ = 2, free space; γ > 2, other media).

The RF coverage map can be obtained from the MTX from a set of RSL(x,y,z). The presence of intelligence in the MTX enables the full automation of this system.

5.11.4 Sampling Rate and Vehicle Speed

In the multipath environment, a pattern of standing waves is generated when the mobile is in motion. This is shown in Figure 5.41 for standing waves generated due to a plane reflecting surface. Since these standing waves repeat every $\lambda/2$, the sampling interval should be $\approx \lambda/4$ or two samples per standing wave. This is known as the Nyquist rate.

Thus, we write:

$$\text{Sampling interval} = \lambda/4 \qquad (5.103)$$

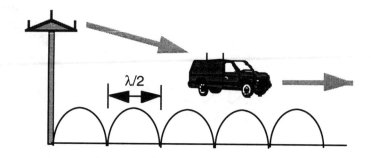

Figure 5.41 Standing waves are generated when the mobile is in motion.

If the velocity of the mobile is defined as V, then the effective sampling rate can be expressed as a function of mobile velocity:

$$\text{Effective sampling rate} = 4V/\lambda \qquad (5.104)$$

where

V = velocity of the mobile, m/s
$\lambda = c/f$ = wavelength
c = velocity of light (3×10^8 m/s)
f = frequency, MHz.

For example, the wavelength for f = 900 MHz is given by

$$\lambda = c/f = 3 \times 10^8 (\text{m/s})/900 \text{ MHz} = 0.333 \text{ meters}$$

If the mobile speed is 100 km/hr (27.77 m/s), the sampling rate will be

$$\text{Effective sampling rate} = 4V/\lambda$$
$$= 4 \times 27.77/0.333 \approx 333 \text{ samples/s}$$

Therefore, for a given frequency, the sampling rate is directly proportional to the vehicle speed.

5.12 CONCLUSIONS

We have examined various anomalies of RF propagation and presented a general overview of various prediction and measurement techniques. Although these predictions and measurement techniques are the foundation of today's cellular services, they suffer from inaccuracies due to user-defined clutter factors. These clutter factors arises due to numerous RF barriers, which vary from place to place. It is practically impossible to accommodate these factors accurately. Cell site location is also a challenging engineering task because of regulations and restrictions imposed on some locations. Therefore cell sites have to be relocated from the predicted location, requiring the best judgment of RF engineers.

Thus we come to the conclusion that propagation prediction is a combination of science, engineering, and art. An experienced RF engineer, willing to compromise between theory and practice, is expected to accomplish the most.

References

[1] Hawking, Stephen, *A Brief History of Time,* New York: Bantam Books, 1988.

[2] Kraus, J. D., *Antennas,* New York: McGraw-Hill Book Company, 1950.

[3] IS-54, "Dual-Mode Mobile Station–Base Station Compatibility Standard," PN-2215, Electronic Industries Association Engineering Department, December 1989.

[4] Lee, William C. Y., *Mobile Cellular Telecommunications Systems,* New York: McGraw-Hill Book Company, 1989.

[5] Mehrotra, Asha, *Cellular Radio, Analog and Digital Systems,* Norwood, MA: Artech House, 1994.

[6] Hata, M., "Empirical Formula for Propagation Loss in Land Mobile Radio Services," *IEEE Trans. on Vehicular Technology,* Vol. VT-29, 1980, pp. 317–325.

[7] Walfisch, J., et al., "A Theoretical Model of UHF Propagation in Urban Environments," *IEEE Trans. on Antennas and Propagation,* Vol. AP-38, 1988, pp. 1788–1796.

[8] Milstein, L. B., et al., "On the Feasibility of a CDMA Overlay for Personal Communications Network," *IEEE J. on Selected Areas in Communications,* Vol.10, No. 4, May 1992, pp. 655–668.

[9] Xia, H. H., et al., "Radio Propagation Characteristics for Line of Sight Microcellular and Personal Communications," *IEEE Trans. on Antennas and Propagation,* Vol. 41, No. 10, October 1993, pp. 1439–1447.

[10] Faruque, Saleh, "PCS Microcells Insensitive to Propagation Medium," *Proc. IEEE Globecom '94,* Vol. 1, 1994, pp. 32–36.

[11] Honcharenko, W., et al., "Mechanisms Governing UHF Propagation on Single Floors in Modern Office Buildings," *IEEE Trans. on Vehicular Technology,* Vol. 41, No. 4, November 1992, pp. 496–504.

[12] Kreyszig, Erwin, *Advanced Engineering Mathematics,* 5th ed., New York: John Wiley & Sons, 1983.

[13] Miller, Irwin, and John E. Freund, *Probability and Statistics for Engineers,* Englewood Cliffs, NJ: Prentice-Hall, 1977.

[14] Kosko, Bart, *Fuzzy Thinking,* New York: Hyperion, 1993.

[15] Kosko, Bart, and Satoru Isaka, "Fuzzy Logic," *Scientific American,* July 1993, pp. 76–81.

[16] Faruque, Saleh, "Propagation Prediction Based on Environmental Classification and Fuzzy Logic Approximation," *Proc. IEEE ICC'96,* 1996, pp. 272–276.

[17] Faruque, Saleh, "An Automated Reverse-Path Radio Survey," *Proc. Wireless'93, Fifth International Conference on Wireless Communications,* Calgary, Canada, 1993, pp. 377–381.

CHAPTER 6

▼▼▼

THE ART OF TRAFFIC ENGINEERING

6.1 INTRODUCTION

Traffic engineering is a branch of science that deals with provisioning of communication circuits in a given service area, for a given number of subscribers, with a given grade of service. It is also a process of revenue prediction. An overprovisioned system (Figure 6.1(a)) guarantees 100% system availability, but it is not cost effective. On

(a)

(b)

Figure 6.1 Conceptual representation of (a) overprovision and (b) underprovision.

the other hand, an underprovisioned system (Figure 6.1(b)) that has too many subscribers per channel, is responsible for call blocking, denying services to many subscribers. A good compromise between these two is where the art of traffic engineering comes into play.

Traffic engineering for a cellular network involves (1) acquiring demographic data, (2) translating demographic data into traffic (erlangs) per square mile or per square kilometer, (3) mapping a hexagonal grid over the traffic distribution pattern in a given service area, (4) assigning an appropriate number of channels per cell, (5) estimating total number of cells, and (6) provisioning switches. This is a challenging task for the designer, requiring a good understanding of traffic distribution, traffic growth, customer requirements, and careful system planning.

This may be best explained by means of the system diagram shown in Figure 6.2 where the cellular traffic is comprised of *land to mobile calls, mobile to land calls*, and *mobile to mobile calls*. The cellular segment of the communication is shown as a cluster of seven hexagonal cells. Note that a seven-cell cluster is a widely used cell plan where all 416 available channels are evenly distributed among seven cells.

Each cell is essentially a radio communication center in which a mobile subscriber establishes a call with a land telephone through the MTX and the PSTN. This composite platform enables us to communicate with anyone at any time from anywhere within the service area. To offer these services, it is necessary to design the cellular network in such a way that it provides satisfactory services to the subscriber with a minimum cost, while generating revenue for the service provider.

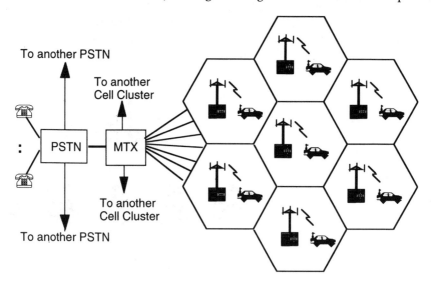

Figure 6.2 Elements of cellular communication system.

The objective of this chapter is to provide a basic understanding of traffic engineering and present engineering aspects of cell site provisioning as practiced in modern cellular communication systems. Frequency planning and cell site engineering also play important roles in determining the overall system traffic and system performance. These engineering tasks stand alone by dint of their own merits and are discussed in Chapter 7 and Chapter 8, respectively.

6.2 TRAFFIC CHARACTERISTICS

All traffic engineering is based on the average busy-hour traffic during a day. This busy hour is generally consistent and, therefore, predictable. A typical distribution pattern of cellular traffic is given in Figure 6.3 [1]. It has two peaks: before noon and before evening. These peaks are known as *busy-hour traffic*. Traffic is generally low during the night and rises rapidly in the morning when offices, shops, and factories open for business. Traffic intensity gradually goes down during lunchtime and picks up again in the afternoon. These variations are commonly known as *hourly variations*. Figure 6.3 also indicates higher average intensities on business days and lower traffic activities during weekends and holidays. These variations are known as *daily variations*.

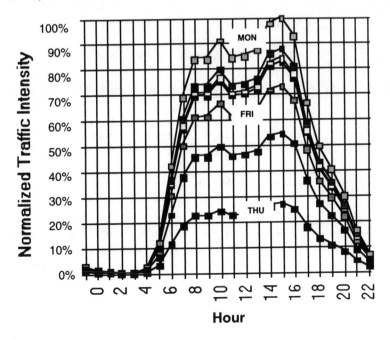

Figure 6.3 A typical distribution pattern of cellular traffic.

There are other variations as well, as described here:

- *Call holding time variations:* Call holding time is the average duration of a call. It varies according to type of subscriber (i.e., business, private, etc.). Typical call holding time varies between 120 and 180 sec. This is an important parameter in estimating the traffic as we shall see later.
- *Seasonal variations:* Around the major holidays, it may be necessary to provision more circuits.
- *Long-term variations:* Gradual subscriber growth over a period of years. These variations have to be taken into account for long-term growth and system planning.

6.3 INTENSITY AND UNITS OF TRAFFIC

Traffic intensity is measured in Erlangs, where 1 Erlang = one circuit in use for 1 hr (3600 sec), named after the Danish mathematician, A. K. Erlang, founder of the theory of telephone traffic. Traffic intensity is also measured in CCS (circuit centum seconds) per hour, where 1 CCS = one circuit in use for 100 sec.

The relationship between Erlangs and CCSs can be defined as follows:

$$1 \text{ Erlang} = 1 \text{ circuit in use for } 3600 \text{ sec}$$

$$1 \text{ CCS} = 1 \text{ circuit in use for } 100 \text{ sec}$$

Therefore, Erlang/CCS = 36 or 1 Erlang = 36 CCS. These traffic units are defined as

$$\text{Erlang} = \frac{\text{Number of calls} \times \text{Average call holding time (sec)}}{3600} \qquad (6.1)$$

$$\text{CCS} = \frac{\text{Number of calls} \times \text{Average call holding time (sec)}}{100} \qquad (6.2)$$

Either Erlangs or CCSs can be used to determine the total number of channels required to meet the service objectives for a given blocking rate. Number of calls is the total number of calls that can be handled for a given Erlang or CCS.

6.4 GRADE OF SERVICE

Grade of service (GOS) is defined as the probability of call failure. It means that a call will be lost due to transmission congestion (i.e., when all the available channels are busy, any additional calls will be denied access to the communication system). GOS lies between 0 and 1:

$$0 < GOS < 1 \tag{6.3}$$

All calls will fail if GOS = 1. (This means no service at all, zero revenue.) All calls will pass if GOS = 0. (This is overprovision, poor revenue.) Typically, GOS = 0.02 in cellular communication systems.

6.5 TRAFFIC CALCULATIONS (POISSON'S DISTRIBUTION)

The Poisson distribution [2,3] is a statistical process that applies to a sequence of events that take place at regular intervals of time or throughout a continuous interval of time. A Poisson distribution has many important applications; for instance, we may be interested in the number of customers arriving for service at a gasoline station, the number of airplanes arriving at an airport, or the number of phone calls arriving at a switch. The mathematical model that describes many situations like these is the Poisson distribution. Let N be the total number of trunks (channels) T be the offered traffic in Erlangs (or CCSs). Then the probability of all the channels being busy will be given by the following Poisson distribution:

$$P(N;T) = \frac{T^N e^{-T}}{N!} \tag{6.4}$$

where $P(N; T)$ is the blocking rate or GOS. Thus, for a given traffic capacity and blocking rate, the number of radios can be calculated. Table 6.1 provides a list of offered traffic in Erlangs as a function of blocking probability. Additional traffic data for a GOS other than 2% may be obtained from elsewhere [4,5]. Note that the most often used table for telephony is the Erlang B. It assumes that blocked calls are cleared and that the caller tries again later. There are other tables such as Erlang C, which assumes that blocked calls are retried until the call is established.

Figure 6.4 shows the relationship between Erlangs and the number of channels for GOS = 2%, which can be used for a quick estimation of traffic for 2% blocking rate.

Example

Given 20 subscribers, each of which generates 0.1 Erlang/h, and four available channels, what is the probability that all four channels will be busy?

Table 6.1
Erlang B Table With GOS = 2%, the Most Frequently Used Percentage for Cellular Applications

#Trunks	Erlangs	#Trunks	Erlangs	#Trunks	Erlangs	#Trunks	Erlangs	#Trunks	Erlangs	#Trunks	Erlangs	#Trunks	Erlangs	#Trunks	Erlangs
1	0.0204	26	18.4	51	41.2	76	64.9	100	88	150	136.8	200	186.2	250	235.8
2	0.223	27	19.3	52	42.1	77	65.8	102	89.9	152	138.8	202	188.1	300	285.7
3	0.602	28	20.2	53	43.1	78	66.8	104	91.9	154	140.7	204	190.1	350	335.7
4	0.109	29	21	54	44	79	67.7	106	93.8	156	142.7	206	192.1	400	385.9
5	1.66	30	21.9	55	44.9	80	68.7	108	95.7	158	144.7	208	194.1	450	436.1
6	2.28	31	22.8	56	45.9	81	69.6	110	97.7	160	146.6	210	196.1	500	486.4
7	2.94	32	23.7	57	46.8	82	70.6	112	99.6	162	148.6	212	198.1	600	587.2
8	3.63	33	24.6	58	47.8	83	71.6	114	101.6	164	150.6	214	200	700	688.2
9	4.34	34	25.5	59	48.7	84	72.5	116	103.5	166	152.6	216	202	800	789.3
10	5.08	35	26.4	60	49.6	85	73.5	118	105.5	168	154.5	218	204	900	890.6
11	5.84	36	27.3	61	50.6	86	74.5	120	107.4	170	156.5	220	206	1000	999.1
12	6.61	37	28.3	62	51.5	87	75.4	122	109.4	172	158.5	222	208	1100	1093
13	7.4	38	29.2	63	52.5	88	76.4	124	111.3	174	160.4	224	210		
14	8.2	39	30.1	64	53.4	89	77.3	126	113.3	176	162.4	226	212		
15	9.01	40	31	65	54.4	90	78.3	128	115.2	178	164.4	228	213.9		
16	9.83	41	31.9	66	55.3	91	79.3	130	117.2	180	166.4	230	215.9		
17	10.7	42	32.8	67	56.3	92	80.2	132	119.1	182	168.3	232	217.9		
18	11.5	43	33.8	68	57.2	93	81.2	134	121.1	184	170.3	234	219.9		
19	12.3	44	34.7	69	58.2	94	82.2	136	123.1	186	172.4	236	221.9		
20	13.2	45	35.6	70	59.1	95	83.1	138	125	188	174.3	238	223.9		
21	14	46	36.5	71	60.1	96	84.1	140	127	190	176.3	240	225.9		
22	14.9	47	37.5	72	61	97	85.1	142	128.9	192	178.2	242	227.9		
23	15.8	48	38.4	73	62	98	86	144	130.9	194	180.2	244	229.9		
24	16.6	49	39.3	74	62.9	99	87	146	132.9	196	182.2	246	231.8		
25	17.5	50	40.3	75	63.9	100	88	148	134.8	198	184.2	248	233.8		

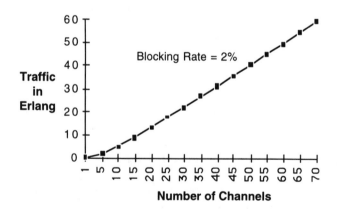

Figure 6.4 Relationship between Erlangs and number of channels.

Answer

$$N = 4$$
$$E = 0.1 \times 20 = 2$$

$$G(4;2) = \frac{2^4 e^{-2}}{4!} = 0.09$$

that is, there is a 9% probability of four channels being busy or GOS = 0.09.

Example

Given the following parameters, determine the total number of channels: number of calls expected = 1000; average call holding time = 120 sec; GOS = 0.02.

Answer

$$\text{Total traffic in Erlangs} = 1000 \times 120/3600 = 33 \text{ Erlangs}$$

From the Erlang B table or from the graph of Figure 6.4, we find that 33 Erlangs at 2% GOS represents approximately 42 channels. Therefore, for all practical purposes, an OMNI site having 48 channels is a good compromise. For an urban area, it may be necessary to add sectors to the cell to combat multipath.

6.6 PRINCIPLE OF BASE STATION PROVISIONING

Base station provisioning (cell site provisioning) is a step-by-step process of assigning a certain number of channels per base station. It involves acquisition of population density, translation of population density into traffic data (Erlangs) per square mile or per square kilometer, mapping a hexagonal grid over the traffic distribution pattern, computing the number of channels per cell, and predicting the total number of cells. These are briefly described in the following subsections.

6.6.1 Acquisition of Population Density

Figure 6.5 shows a typical traffic distribution pattern over a given geographic area. Each square (termed a *bin*) represents population density, translated into Erlangs and using the Erlang formula of (6.1). The total bin count then represents the offered traffic in a given service area.

6.6.2 Hexagonal Cell Grid Overlay

The next step of the process is to superimpose a hexagonal cell grid over the entire service area. This is shown in Figure 6.6 where the traffic per cell is determined by

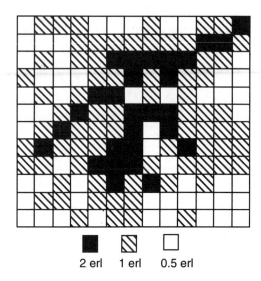

2 erl   1 erl   0.5 erl

Figure 6.5 Traffic distribution pattern over a service area of interest. Population density is indicated by unit square, translated into Erlangs.

tallying the bin count in each cell. For example a cell may have 4 bins with 2 Erlangs, 5 bins with 1 Erlang, and 2 bins with 0.5 Erlang. The total estimated traffic in that cell would be $(4 \times 2) + (5 \times 1) + (2 \times 0.5) = 14$ Erlangs. With 2% GOS, this translates into 21 channels or 21 cell site radios. Therefore, the purpose of a hexagonal grid overlay is to obtain bin count per cell, which in turn determines the number of channels for a given grade of service.

6.6.3 Traffic Bin Count

Bin count is a measure of traffic in a given cell site. Thus, referring to Figure 6.6, we find that there are three different types of bins, namely, 2-Erlang bins, a 1-Erlang bin, and a 0.5-Erlang bin. For convenience we represent them as (n, m, k) bins where 2-Erlang bins are represented by n, 1-Erlang bins are represented by m, and 0.5-Erlang bins are represented by k. For example, a cell having four 2-Erlang bins, five 1-Erlang bins, and two 0.5-Erlang bins would be represented by $(4,5,2)$. In other words the bin count of the cell is $(4, 5, 2)$. Figure 6.7 shows the entire hexagonal grid tallied with bin counts. These bin counts are proportional to the population density, which forms the basis of cell site provisioning.

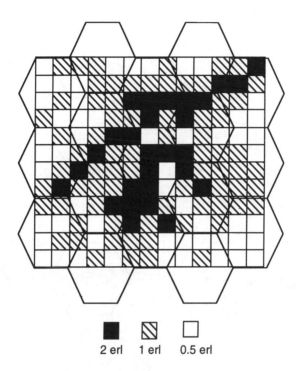

2 erl   1 erl   0.5 erl

Figure 6.6 Superposition of hexagonal grid over the traffic pattern.

6.6.4 Traffic Count in Erlangs

Traffic count in Erlangs is derived from the bin count, described in section. For example, the traffic in erlang corresponding to a bin cc given by $(4 \times 2) + (5 \times 1) + (2 \times 0.5) = 14$ Erlangs. Repeating this proc the grid, we obtain the traffic count per cell as shown in Figure 6.'

6.6.5 Channel Count

Once the traffic count is determined in terms of Erlangs, t' can be determined by using the Erlang B table. Because ' station radios, it is customary to use a high GOS. A GO' for cellular base stations. This is shown in Figure 6.9, whi' to an end.

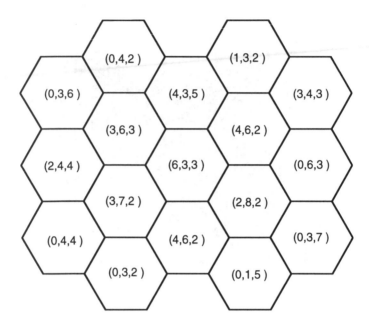

Figure 6.7 Traffic bin count per cell.

6.7 DESIGN CONSTRAINTS

Several considerations have to be made at this point. The foremost consideration is the frequency planning, which determines the channel capacity per cell. For example, the $N = 7$ frequency plan provides $416/7 = 48$ channels per cell (see Chapter 7). On the other hand, the $N = 4$ frequency plan provides $416/4 = 104$ channels per cell or $104/3 \approx 34$ channels per sector in a tricellular plan. Therefore, if the bin count, presented earlier, yields more than 48 channels per cell, then the $N = 7$ frequency plan cannot be used. However, this frequency plan could still be used if the cell is reduced in size. But this poses the problem of requiring more cells to accommodate the bin counts in a given service area. Alternately, the $N = 4$ frequency plan, which has a higher channel count, could be used. Cell site location and a choice between OMNI and sectorization are among the other design constraints the designer consider in order to balance these contradictory issues.

TRUNKING EFFICIENCY

Trunking efficiency (channel utilization efficiency) is also known as a measure of concentration efficiency. It is determined by the amount of traffic per channel, defined

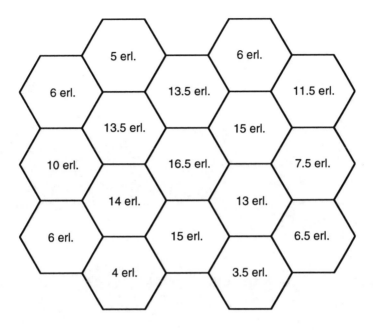

Figure 6.8 Erlangs per cell.

$$\text{Efficiency } (\%) = \frac{\text{Traffic in Erlangs}}{\text{Number of channels}} \times 100 \qquad (6.5)$$

From the Erlang table with GOS = 2%, we obtain the graph of Figure 6.10, which shows the relationship between efficiency and capacity [6]. We see that for a given GOS, the efficiency increases as the number of trunks (voice circuits) increases. A cell site having fewer than 15 voice circuits (channels) is generally inefficient, less cost effective, and generates poor revenue.

To illustrate this further, we consider an OMNI cell having 48 channels, which are then sectored into three as shown in Figure 6.11. Although the total number of channels in the sectored cell remains the same, the traffic capacity of the OMNI cell is higher than the sectored cell due to trunking efficiency.

Example

For the Omni site:

Total number of channels available	=	48
GOS	=	0.02
Then the traffic intensity will be	=	38.4 Erlangs
Therefore, trunking efficiency	=	38.4/48 = 0.8 (80%).

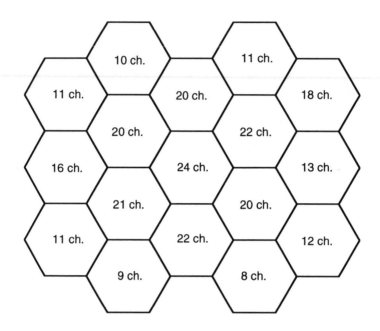

Figure 6.9 Channel count.

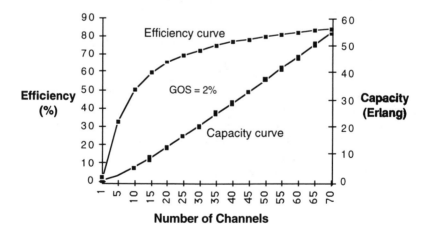

Figure 6.10 Trunking efficiency.

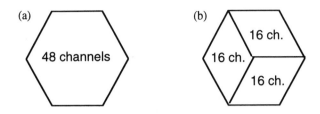

Figure 6.11 Illustration of trunking efficiency. (a) OMNI cell. (b) Sectored cell.

For the sectored site:

The available 48 channels are divided into three groups, with 16 channels per group.

Number of channels/group	=	16
GOS	=	0.02
Then the traffic intensity/group will be	=	9.83 Erlangs
Total traffic in three sectors (one cell)	=	29.49 Erlangs
Therefore, trunking efficiency	=	29.49/48 = 0.614 (61.4%).

References

[1] Faruque, Saleh, and Scott Baxter, "RF Engineering Seminar 1000," An internal RF engineering course, Northern Telecom, Richarson, Texas, 1994.

[2] Kreyszig, Erwin, *Advanced Engineering Mathematics*, 5th Ed., New York: John Wiley & Sons, 1983.

[3] Miller, Irwin, and John E. Freund, *Probability and Statistics for Engineers*, Englewood Cliffs, NJ: Prentice-Hall, 1977.

[4] Lee, William C. Y., *Mobile Cellular Telecommunications Systems*, New York: McGraw-Hill Book Company, 1989.

[5] Mehrotra, Asha, *Cellular Radio, Analog and Digital Systems*, Norwood, MA: Artech House, 1994.

[6] Faruque, Saleh, "A Journey into the Realm of Wireless Communication," An internal RF engineering course, Northern Telecom, Brampton, Ontario, Canada, 1993.

CHAPTER 7
▼▼▼

FREQUENCY PLANNING

7.1 INTRODUCTION

Frequency planning is a means to optimize spectrum usage, enhance channel capacity, and reduce interference. The use of a spectrum and licenses to operate radio transmitters is controlled by a number of regulatory authorities:

- Federal communications Commission (FCC) for commercial carriers;
- National Telecommunications and Information Administration (NTIA) for government systems;
- Military Communications and Electronics (MCEB) for military systems;
- Consultative Committee for International Radio (CCIR) for issues and recommendations on radio channel assignment. They do not have regulatory power but their recommendations are usually adopted worldwide.

The FCC provides licenses to operate cellular communication systems over a given band of frequencies. Because cellular communication is a multiple-access system, operators have to comply with the regulations. Compliance requires proper frequency planning and spectrum control and also involves channel numbering, channel grouping into subsets, cell planning, and channel assignment. A frequency plan must ensure adequate channel isolation to avoid energy slipover between channels, so that adjacent channel interference is reduced to a minimum. Moreover, an

---

199

adequate repeat distance should be provided to an extent where cochannel interference is acceptable, but a high channel capacity is maintained. Frequency planning is an important task in cellular communications for system planning, installation, performance analysis, capacity analysis, etc.

To provide a comprehensive overview, this chapter begins with the basic concept of frequency planning. The step-by-step processes involved in cell planning and channel assignment are presented next. Examples of various frequency plans with respect to OMNI and sectored cell sites are given. The associated channel capacity and C/I performances are also evaluated.

7.2 TRANSMITTING/RECEIVING CHANNEL PAIRING

Cellular communication is a full-duplex system in which each band is divided into two equal halves: one-half is for forward channels (base to mobile communication) and the other half is for the reverse channels (mobile to base communication). A 45-MHz guardband is provided to avoid interference between Tx/Rx channels. This is illustrated in Figure 7.1 where N is the channel number.

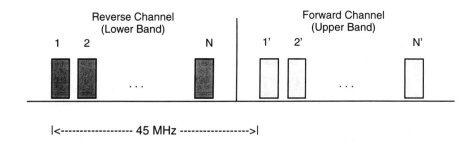

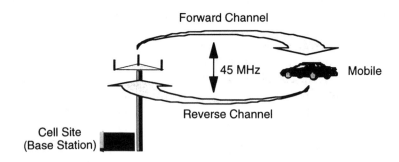

Figure 7.1 Transmitting/receiving channel pairing.

7.3 FREQUENCY BAND ALLOCATION AND CHANNEL NUMBERING

Frequency planning for cellular communications begins with the available frequency band allocated by the regulating authority. This is shown in Figure 7.2 for Band A and Band B. Each band occupies 12.5 MHz, of which 10 MHz is in the nonexpanded spectrum (NES) and 2.5 MHz is in the expanded spectrum (ES) [1].

With 30-kHz channel spacing, this arrangement provides 10 MHz/30 kHz = 333 channels per band in the NES and 2.5 MHz/30 kHz = 83 channels per band in the ES. The total number of channels per band is therefore 333 + 83 = 416 in each band. Among them, 21 channels are used as control channels, and the remaining 395 channels are used as voice channels, as summarized in Table 7.1. Note that the control channels are used for channel assignment, paging, messaging, etc.

Frequency assignment is related to the channel number as follows:

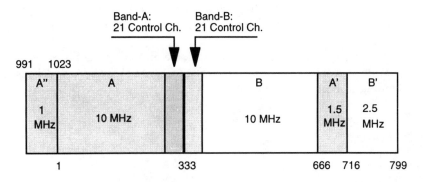

Figure 7.2 Allocated bands for 800-MHz cellular system.

Table 7.1
Channel Numbers

System	Type	BW (MHz)	Number of Channels
A″	ES	1	33
A	NES	10	333
B	NES	10	333
A′	ES	1.5	50
B′	ES	2.5	83

$$\text{Base transmitting frequency} = (0.03N + 870) \text{ MHz} \qquad \text{(NES)}$$
$$= 0.03(N - 1023) + 870 \text{ MHz} \qquad \text{(ES)} \qquad (7.1)$$

$$\text{Base receiving frequency} = 0.03N + 825 \text{ MHz} \qquad \text{(NES)}$$
$$= 0.03(N - 1023) + 825 \text{ MHz} \qquad \text{(ES)} \qquad (7.2)$$

where N is the channel number ($N = 1, 2, \ldots, 1023$). Thus by knowing the channel number, the associated pair of frequencies can be obtained from (7.1) and (7.2).

7.4 CHANNEL GROUPING AND REUSE PLAN

Cellular communication is a multiple-access system in which several noninterfering channels are combined to form a channel group and assigned to a cell site. Because the number of channels is limited, these channel groups are reused at a regular distances. This is an important engineering task because it determines system capacity and performance.

Several frequency reuse techniques are available and are generally known as *frequency planning* or *channel assignment* techniques. Some of the most widely used frequency planning techniques are as follows:

- $N = 7$ frequency reuse plan;
- $N = 9$ frequency reuse plan;
- $N = 4$ frequency reuse plan;
- $N = 3$ frequency reuse plan.

The $N = 3$ and $N = 9$ frequency plans are relatively unknown. However, since the advent of TDMA techniques, these frequency plans are becoming popular among cellular industries for enhancing channel capacity.

7.5 $N = 7$ OMNI FREQUENCY PLAN

In the $N = 7$ frequency reuse plan, the available channels are equally divided among seven cells known as a *seven-cell cluster*. Because there are 21 control channels, this frequency plan is based on dividing all the available frequencies into 21 frequency groups (see Tables 7.2(a) and (b)) while one control channel is assigned per group [1]. The corresponding $N = 7$ cell cluster, with three frequency groups per cell, is shown in Figure 7.3.

The total number of frequency groups per cluster is $7 \times 3 = 21$. Channel assignment is based on the following sequence: ($N, N + 7, N + 14$) where N is the cell number ($N = 1, 2, \ldots, 7$). The channel grouping scheme is shown in Table 7.3 and the corresponding $N = 7$ OMNI cell cluster is shown in Figure 7.3.

Table 7.2(a)
N = 7/21 Band A Frequency Chart

1	2	3	4	5	6	7	8	9	10	11	12	13	14	15	16	17	18	19	20	21
22	23	24	25	26	27	28	29	30	31	32	33	34	35	36	37	38	39	40	41	42
43	44	45	46	47	48	49	50	51	52	53	54	55	56	57	58	59	60	61	62	63
64	65	66	67	68	69	70	71	72	73	74	75	76	77	78	79	80	81	82	83	84
85	86	87	88	89	90	91	92	93	94	95	96	97	98	99	100	101	102	103	104	105
106	107	108	109	110	111	112	113	114	115	116	117	118	119	120	121	122	123	124	125	126
127	128	129	130	131	132	133	134	135	136	137	138	139	140	141	142	143	144	145	146	147
148	149	150	151	152	153	154	155	156	157	158	159	160	161	162	163	164	165	166	167	168
169	170	171	172	173	174	175	176	177	178	179	180	181	182	183	184	185	186	187	188	189
190	191	192	193	194	195	196	197	198	199	200	201	202	203	204	205	206	207	208	209	210
211	212	213	214	215	216	217	218	219	220	221	222	223	224	225	226	227	228	229	230	231
232	233	234	235	236	237	238	239	240	241	242	243	244	245	246	247	248	249	250	251	252
253	254	255	256	257	258	259	260	261	262	263	264	265	266	267	268	269	270	271	272	273
274	275	276	277	278	279	280	281	282	283	284	285	286	287	288	289	290	291	292	293	294
295	296	297	298	299	300	301	302	303	304	305	306	307	308	309	310	311	312				
313	**314**	**315**	**316**	**317**	**318**	**319**	**320**	**321**	**322**	**323**	**324**	**325**	**326**	**327**	**328**	**329**	**330**	**331**	**332**	**333**
															667	668	669	670	671	672
673	674	675	676	677	678	679	680	681	682	683	684	685	686	687	688	689	690	691	692	693
694	695	696	697	698	699	700	701	702	703	704	705	706	707	708	709	710	711	712	713	714
715	716																			
									991	992	993	994	995	996	997	998	999	1000	1001	1002
1003	1004	1005	1006	1007	1008	1009	1010	1011	1012	1013	1014	1015	1016	1017	1018	1019	1020	1021	1022	1023

This scheme provides 21 × 30 kHz = 630-kHz channel isolation within a cell and no isolation between cells [see Fig. 7.3(a)]. In Figure 7.3(b) an optimized version of N = 7 plan is given that provides fewer adjacent channels due to channel reassignment within the cluster. However, channel adjacency reappears due to frequency reuse as shown in Figure 7.4.

The preceding analysis indicates that the total elimination of channel adjacency is practically impossible in the N = 7 plan, which gives rise to adjacent channel interference throughout the network.

7.5.1 Evaluation of Cochannel Interference

A cochannel interferer has the same nominal frequency as the desired frequency. It arises from multiple uses of the same frequency, which may be expressed in a number of ways:

Table 7.2(b)
$N = 7/21$ Band B Frequency Chart

334	335	336	337	338	339	340	341	342	343	344	345	346	347	348	349	350	351	352	353	354
355	356	357	358	359	360	361	362	363	364	365	366	367	368	369	370	371	372	373	374	375
376	377	378	379	380	381	382	383	384	385	386	387	388	389	390	391	392	393	394	395	396
397	398	399	400	401	402	403	404	405	406	407	408	409	410	411	412	413	414	415	416	417
418	419	420	421	422	423	424	425	426	427	428	429	430	431	432	433	434	435	436	437	438
439	440	441	442	443	444	445	446	447	448	449	450	451	452	453	454	455	456	457	458	459
460	461	462	463	464	465	466	467	468	469	470	471	472	473	474	475	476	477	478	479	480
481	482	483	484	485	486	487	488	489	490	491	492	493	494	495	496	497	498	499	500	501
502	503	504	505	506	507	508	509	510	511	512	513	514	515	516	517	518	519	520	521	522
523	524	525	526	527	528	529	530	531	532	533	534	535	536	537	538	539	540	541	542	543
544	545	546	547	548	549	550	551	552	553	554	555	556	557	558	559	560	561	562	563	564
565	566	567	568	569	570	571	572	573	574	575	576	577	578	579	580	581	582	583	584	585
586	587	588	589	590	591	592	593	594	595	596	597	598	599	600	601	602	603	604	605	606
607	608	609	610	611	612	613	614	615	616	617	618	619	620	621	622	623	624	625	626	627
628	629	630	631	632	633	634	635	636	637	638	639	640	641	642	643	644	645	646	647	648
649	650	651	652	653	654	655	656	657	658	659	660	661	662	663	664	665	666			
					717	718	719	720	721	722	723	724	725	726	727	728	729	730	731	732
733	734	735	736	737	738	739	740	741	742	743	744	745	746	747	748	749	750	751	752	753
754	755	756	757	758	759	760	761	762	763	764	765	766	767	768	769	770	771	772	773	774
775	776	777	778	779	780	781	782	783	784	785	786	787	788	789	790	791	792	793	794	795
796	797	798	799																	

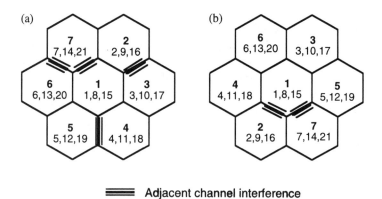

Figure 7.3 (a) $N = 7$ OMNI frequency reuse plan showing adjacent channel interference. (b) $N = 7$ OMNI frequency reuse plan showing reduced adjacent channel interference within the cluster.

Table 7.3
N = 7/21 OMNI Channel Grouping Scheme

Cell Number	1	2	3	4	5	6	7
Frequency Group:							
N	1	2	3	4	5	6	7
N + 7	8	9	10	11	12	13	14
N + 14	15	16	17	18	19	20	21

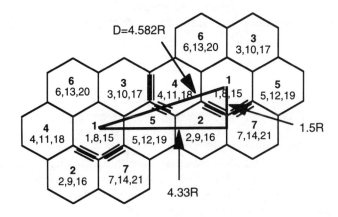

Figure 7.4 N = 7 OMNI growth plan showing reappearance of adjacent channels.

1. Cochannel Interference (CCI) or interference-to-carrier ratio (I/C);
2. Carrier-to-interference ratio (C/I).

The difference between CCI and C/I is then CCI (dB) = −C/I (dB). For OMNI sites (see Figure 7.4) this is given by [2]

$$\frac{C}{I} = 10 \log\left[\frac{1}{j}\left(\frac{D}{R}\right)^{\gamma}\right] \tag{7.3}$$

where

j	=	number of cochannel interferers ($j = 1, 2, \ldots, 6$)
γ	=	propagation constant
D	=	frequency reuse distance
R	=	cell radius.

The distance ratio (reuse distance) D/R is given by [2–4]

$$\frac{D}{R} = \sqrt{3N} \tag{7.4}$$

For the $N = 7$ frequency plan,

$$\frac{D}{R} = \sqrt{3 \times 7} = 4.58 \tag{7.5}$$

The distance ratio can also be determined geometrically as shown in Figure 7.4.

With $\gamma = 4$, repeat distance = 4.58, and $j = 6$, the carrier-to-interference ratio becomes

$$C/I = 18.6 \text{ dB} \qquad \text{or} \qquad CCI = -18 \text{ dB} \tag{7.6}$$

7.5.2 Evaluation of Adjacent Channel Interference

Adjacent channel interference arises from energy slipover between two adjacent channels. This can be evaluated with the aid of Figure 7.5 where it is assumed that the adjacent channel is assigned to the adjacent site and the ratio d_i/d_c varies as the mobile moves toward or away from the cell. Moreover, the out-of-band signals are also assumed attenuated by the postmodulation filter by at least 26 dB (EIA standard) [1]. Then the adjacent channel interference will be:

$$ACI = -10 \log\left[\left(\frac{d_i}{d_c}\right)^{\gamma}\right] + \text{Adjacent channel isolation} \tag{7.7}$$

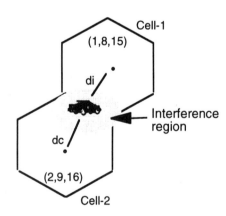

Figure 7.5 Adjacent channel interference evaluation scheme.

where the adjacent channel isolation is provided by the postmodulation filter, which is generally ≈ 26 dB. As an example, an adjacent channel in the adjacent site has $d_i/d_c = 1$, resulting in ACI ≈ -26 dB. An adjacent channel in an alternate channel (one cell away) provides much greater than 26-dB isolation due to a higher distance ratio ($d_i/d_c > 1$).

7.5.3 Channel Capacity in $N = 7$ OMNI Frequency Plan

Channel capacity is measured by the available voice channels per cell, translated into Erlangs (see Chapter 6). Because there are 21 control channels, the number of voice channels is $333 - 21 = 312$ in the NES and $416 - 21 = 395$ in the ES. The channel capacity per cell can be evaluated as:

$$\text{NES channel capacity} = 312/7 \approx 44 \text{ voice channels per cell}$$
$$= 35 \text{ Erlangs @ 2\% GOS} \qquad (7.8)$$

Similarly for the ES, the channel capacity becomes

$$\text{ES channel capacity} = 395/7 = 56 \text{ channels per cell}$$
$$= 46 \text{ Erlangs @ 2\% GOS} \qquad (7.9)$$

In (7.8) and (7.9), GOS is the grade of service or call blocking rate where $0 < \text{GOS} < 1$. 2% grade of service is generally used in North American telephony.

Table 7.4 summarizes these results for several different GOS values.

7.5.4 $N = 7$ Trapezoidal Plan With Alternate Channel Assignment

A trapezoidal plan is based on arranging a cluster of seven cells in two rows that have alternating channel assignments. This is shown in Figure 7.6(a) where the adjacent channels are completely eliminated. Because of the appearance of the cluster, this configuration is termed a *trapezoidal* configuration.

Table 7.4
$N = 7/21$ Channel Capacity per OMNI Cell

Number of Channels per Cell		Channel Capacity per Cell in Erlangs			
		GOS = 1%	GOS = 2%	GOS = 3%	GOS = 5%
NES	44	32.5	35	37	39.6
ES	56	42.3	46	48	50

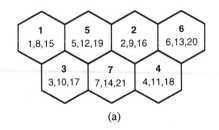

(a)

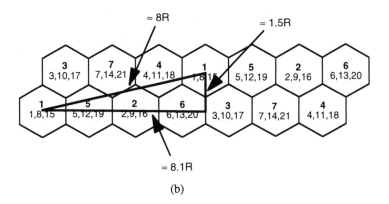

(b)

Figure 7.6 (a) $N = 7$ trapezoidal plan and (b) its horizontal growth plan.

The growth plan is given in Figure 7.6(b). Only horizontal expansion is possible in this plan; for this reason it is suitable for highways and coastal areas.

The reuse distance may be obtained by means of plane geometry. This is given by $D/R = 8.1$ where the cochannel interference, with $\gamma = 4$ is

$$C/I = 10 \log[0.5(8.1)^4] = 33.3 \text{ dB} \tag{7.10}$$

7.6 $N = 9$ OMNI FREQUENCY PLAN

Interference is a major concern in cellular communication systems. It is generally determined by cochannel interference, which in turn depends on frequency planning and antenna engineering. The most popular frequency plan known today is the $N = 7$ plan, which is also known to exhibit adjacent channel interference.

Although the $N = 9$ frequency plan has been around for several years, its practical application had been overlooked by cellular industries. Recently, a frequency assignment technique was proposed [5] for $N = 9$ and $N = 12$ systems to minimize complexity and enhance traffic capacity. In this section, we present an optimized $N = 9$ frequency plan [6], which avoids adjacent channels from adjacent sites. We show that the proposed frequency plan outperforms the existing $N = 7$ frequency plan and is well suited for TDMA digital cellular.

7.6.1 C/I Considerations

The cochannel interference in urban environment is given by

$$\frac{C}{I} = 10 \log\left\{\frac{1}{6}\left[\left(\frac{D}{R}\right)^{\gamma}\right]\right\} \qquad \text{(assuming six interferers)} \qquad (7.11)$$

Solving for the distance ratio, we obtain:

$$\left(\frac{D}{R}\right)^{\gamma} = 6\left[10^{\frac{1}{10}\left(\frac{C}{I}\right)}\right] \qquad (7.12)$$

with $C/I = 17$ dB (EIA standard); the minimum distance ratio becomes $D/R \approx 5$. This means that a given cell cluster must ensure a repeat distance ≥ 5. This is illustrated in Figure 7.7 where an array of four cells is shown, the fourth being a repeated cell. This indicates that a cell cluster must be formed by means of a 3×3 array of cells, resulting in the $N = 9$ pattern.

7.6.2 ACI Considerations

Adjacent channel interference (ACI) occurs because of energy slipover between adjacent channels. It can be minimized or removed by removing adjacent channels and reassigning them elsewhere. This can be achieved by means of the following alternate channel assignment scheme:

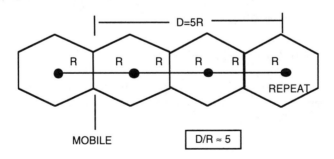

Figure 7.7 Optimum frequency repeat distance evaluation model.

$$1, 3, 5, 7, 9, \ldots, 2, 4, 6, 8, \text{etc.}$$

which ensures adjacent channel isolation. It follows that the cell cluster must be formed by means of a 3×3 array of cells having alternate channel assignment.

7.6.3 The $N = 9$ OMNI Plan

The $N = 9$ frequency plan is based on dividing all the available channels into nine or multiples of nine frequency groups. This is shown in Table 7.5(a) for Band A and in Table 7.5(b) for the Band B. This plan provides 333/18 = 18 channels per group in the NES and 416/18 = 23 channels in the ES. Channel isolation within a group is 18×30 kHz = 540 kHz, which is generally adequate for cellular application. However, the channel isolation can be increased to $540 \times 2 = 1080$ kHz by means of alternate channel assignment before combining (see Chapter 8).

The basic $N = 9$ cell cluster uses an alternate channel distribution scheme which forms a 3×3 array of nine frequency groups. This is shown in Figure 7.8 along with the $N = 9$ cell plan. The rhombic pattern of the cluster is due to the hexagonal pattern of the cell. Unlike the $N = 7$ frequency plan, this frequency plan does not have adjacent channels within the cell cluster. Hence, the ACI performance of this plan is better than any of the existing plans and is well suited for TDMA digital transition.

7.6.4 $N = 9$ OMNI Growth Plan

Because of a rhombic pattern, only vertical and horizontal expansions are possible in this scheme. This is based on repeating the vertical and horizontal sequences shown in Figure 7.9 along with the cell plan.

Since each cell can be visualized as being the center cell of six surrounding cells, a given $N = 7$ frequency plan can be easily translated into a $N = 9$ frequency plan without cell site relocation (see Figure 7.9). This can be accomplished simply by means of frequency reassignment.

7.6.5 Evaluation of Cochannel Interference

Cochannel interference arises from frequency reuse. This can be computed by means of frequency repeat distances associated with this scheme. Thus, referring to Figure 7.9, four cochannel interferers from a repeat distance of

$$D/R \approx (3 \times 9)^{1/2} \approx 5.2$$

Thus the cochannel interference can be obtained as:

Table 7.5(a)
N = 9/18 Band A Frequency Chart

1	2	3	4	5	6	7	8	9	10	11	12	13	14	15	16	17	18
1	2	3	4	5	6	7	8	9	10	11	12	13	14	15	16	17	18
19	20	21	22	23	24	25	26	27	28	29	30	31	32	33	34	35	36
37	38	39	40	41	42	43	44	45	46	47	48	49	50	51	52	53	54
55	56	57	58	59	60	61	62	63	64	65	66	67	68	69	70	71	72
73	74	75	76	77	78	79	80	81	82	83	84	85	86	87	88	89	90
91	92	93	94	95	96	97	98	99	100	101	102	103	104	105	106	107	108
109	110	111	112	113	114	115	116	117	118	119	120	121	122	123	124	125	126
127	128	129	130	131	132	133	134	135	136	137	138	139	140	141	142	143	144
145	146	147	148	149	150	151	152	153	154	155	156	157	158	159	160	161	162
163	164	165	166	167	168	169	170	171	172	173	174	175	176	177	178	179	180
181	182	183	184	185	186	187	188	189	190	191	192	193	194	195	196	197	198
199	200	201	202	203	204	205	206	207	208	209	210	211	212	213	214	215	216
217	218	219	220	221	222	223	224	225	226	227	228	229	230	231	232	233	234
235	236	237	238	239	240	241	242	243	244	245	246	247	248	249	250	251	252
253	254	255	256	257	258	259	260	261	262	263	264	265	266	267	268	269	270
271	272	273	274	275	276	277	278	279	280	281	282	283	284	285	286	287	288
289	290	291	292	293	294	295	296	297	298	299	300	301	302	303	304	305	306
307	308	309	310	311	312	313	314	315	316	317	318	319	320	321	322	323	324
325	326	327	328	329	330	331	332	333									
									667	668	669	670	671	672	673	674	675
676	677	678	679	680	681	682	683	684	685	686	687	688	689	690	691	692	693
694	695	696	697	698	699	700	701	702	703	704	705	706	707	708	709	710	711
712	713	714	715	716													
					991	992	993	994	995	996	997	998	999	1000	1001	1002	1003
1004	1005	1006	1007	1008	1009	1010	1011	1012	1013	1014	1015	1016	1017	1018	1019	1020	1021
1022	1023																

$$C/I = 10 \log[0.25(5.2)^\gamma] \qquad (7.13)$$

With $\gamma = 4$ we obtain:

$$C/I = 22.6 \text{ dB} \qquad \text{(four interferers)} \qquad (7.14)$$

7.6.6 Evaluation of Adjacent Channel Interference

The nearest adjacent channel in $N = 9$ is approximately $3R$ away from the serving base, where R is the cell radius. Thus ACI can be evaluated by means of the following equation:

Table 7.5(b)
N = 9/18 Band B Frequency Chart

334	335	336	337	338	339	340	341	342	343	344	345	346	347	348	349	350	351
352	353	354															
355	356	357	358	359	360	361	362	363	364	365	366	367	368	369	370	371	372
373	374	375	376	377	378	379	380	381	382	383	384	385	386	387	388	389	390
391	392	393	394	395	396	397	398	399	400	401	402	403	404	405	406	407	408
409	410	411	412	413	414	415	416	417	418	419	420	421	422	423	424	425	426
427	428	429	430	431	432	433	434	435	436	437	438	439	440	441	442	443	444
445	446	447	448	449	450	451	452	453	454	455	456	457	458	459	460	461	462
463	464	465	466	467	468	469	470	471	472	473	474	475	476	477	478	479	480
481	482	483	484	485	486	487	488	489	490	491	492	493	494	495	496	497	498
499	500	501	502	503	504	505	506	507	508	509	510	511	512	513	514	515	516
517	518	519	520	521	522	523	524	525	526	527	528	529	530	531	532	533	534
535	536	537	538	539	540	541	542	543	544	545	546	547	548	549	550	551	552
553	554	555	556	557	558	559	560	561	562	563	564	565	566	567	568	569	570
571	572	573	574	575	576	577	578	579	580	581	582	583	584	585	586	587	588
589	590	591	592	593	594	595	596	597	598	599	600	601	602	603	604	605	606
607	608	609	610	611	612	613	614	615	616	617	618	619	620	621	622	623	624
625	626	627	628	629	630	631	632	633	634	635	636	637	638	639	640	641	642
643	644	645	646	647	648	649	650	651	652	653	654	655	656	657	658	659	660
661	662	663	664	665	666												
						717	718	719	720	721	722	723	724	725	726	727	728
729	730	731	732	733	734	735	736	737	738	739	740	741	742	743	744	745	746
747	748	749	750	751	752	753	754	755	756	757	758	759	760	761	762	763	764
765	766	767	768	769	770	771	772	773	774	775	776	777	778	779	780	781	782
783	784	785	786	787	788	789	790	791	792	793	794	795	796	797	798	799	

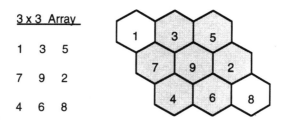

3 x 3 Array

1 3 5

7 9 2

4 6 8

Figure 7.8 Alternate channel assignment in N = 9 OMNI plan.

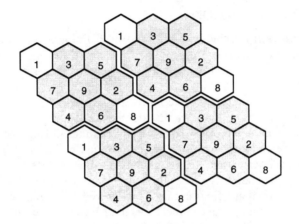

Figure 7.9 $N = 9$ OMNI growth plan.

$$ACI = -[10 \log\{(d_c/d_i)^{-\gamma}\} + \text{Attenuation by radio } (\geq 26 \text{ dB})] \quad (7.15)$$

In the hand-off region, $d_i/d_c \approx 2$. With $\gamma = 4$ this translates into:

$$ACI \leq -38 \text{ dB} \quad (7.16)$$

which is lower than that of the $N = 7$ plan.

7.6.7 $N = 9$ OMNI Capacity

There are 18 frequency groups in the $N = 9$ frequency plan, as shown in Table 7.5. These frequency groups are evenly distributed among nine cells, 2 frequency groups per cell. Out of 21 control channels, only 18 are needed in this scheme. The remaining 3 control channels can be used as voice channels. The number of voice channels is $(333 - 18) = 315$ in the NES and $(416 - 18) = 398$ in the ES. Then the channel capacity can be computed as:

$$\text{NES channel capacity} = 315/9 \approx 35 \text{ voice channels per cell}$$
$$= 26.4 \text{ Erlangs @ 2\% GOS} \tag{7.17}$$

Similarly with ES, the channel capacity becomes

$$\text{ES channel capacity} = 398/9 = 44 \text{ channels per cell}$$
$$= 34.7 \text{ Erlangs @ 2\% GOS} \tag{7.18}$$

Table 7.6 summarizes these results for several different GOS values.

7.7 120-DEG SECTORIZATION

The 120-deg sectorization scheme is achieved by dividing a cell into three sectors of 120 deg each, as shown in Figure 7.10(a). Each sector is treated as a logical OMNI cell where directional antennas are used in each sector for a total of three antennas per cell.

Figure 7.10(b) shows an alternate representation known as the *tricellular plan* [7]. Both configurations are conceptually identical, but the latter is convenient for channel assignment. Each sector uses one control channel and a set of different voice channels. Adequate channel isolations are maintained within and between sectors in order to minimize interference. This is attributed to channel assignment techniques, as we see later in this chapter.

Because directional antennas are used in sectored cells, channels can be reused more frequently, thus enhancing channel capacity. Moreover, multipath components are also reduced due to directionalization, hence enhancing the performance.

7.7.1 $N = 7/21$, 120-Deg Sectored Plan

The $N = 7/21$, 120-deg sectorization plan is based on a distribution of one frequency group per sector with three frequency groups per cell for a total of 21 frequency groups per cluster. This is shown in Figure 7.11(a) for the conventional plan and in Figure 7.11(b) for the tricellular plan where the sector is represented by a hexagon [7]. Channel distribution is based on $(N, N + 7, N + 14)$ where $N = 1, 2, \ldots, 7$. Therefore for $N = 1$, cell 1 uses frequency group 1 for sector 1, frequency group 8 for sector 2, and frequency group 15 for sector 3. Similarly, cell 2 uses frequency groups 2, 9, and 16 for sectors 1, 2, and 3, respectively.

Table 7.6
$N = 9/18$ Channel Capacity per OMNI Cell

Number of Channels per Cell		Channel Capacity per Cell in Erlangs			
		GOS = 1%	GOS = 2%	GOS = 3%	GOS = 5%
NES	35	24.6	26.4	27.7	29.7
ES	44	32.5	34.7	36.2	38.6

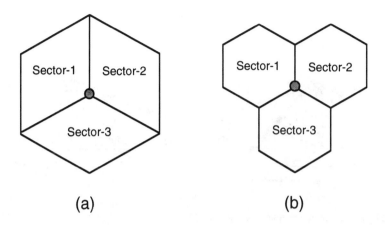

Figure 7.10 Three sector configurations of 120 degrees each. Directional antennas are used in each sector. (a) Conventional representation and (b) an alternate representation.

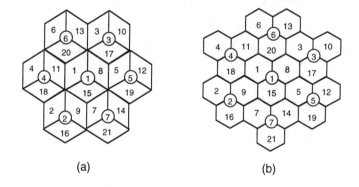

Figure 7.11 (a) $N = 7/21$ sectored configurations based on a 120-deg antenna beam width. (b) $N = 7/21$ tricellular plan using a 60-deg antenna beam width.

7.7.2 $N = 7/21$, 120-Deg Cochannel Interference

The 120-deg sectorization is achieved by dividing a cell into three sectors with directional antennas used in each sector. Thus antenna configuration and their directivity play an important role in determining the C/I performances. To illustrate this further, let us consider the diagram shown in Figure 7.12, in which directional antennas are used for the present analysis. Antenna downtilt is also provided for additional isolation, which must be taken into account.

These assumptions modify the C/I prediction equation as

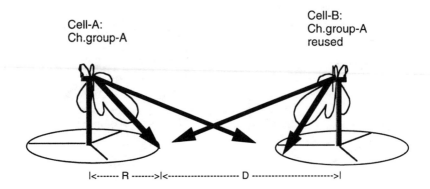

Figure 7.12 Illustration of antenna downtilt and C/I.

$$\frac{C}{I} = 10 \, \log\left[\frac{1}{j}\left(\frac{D}{R}\right)^{\gamma}\right] + \Delta \text{dB} \qquad \text{(due to antenna downtilt)} \qquad (7.19)$$

With $j = 6$, $\gamma = 4$, $D/R = 4.58$, and $\Delta \text{dB} \approx 6$ dB, we now obtain

$$C/I = 24 \text{ dB} \qquad (7.20)$$

The performance can be further improved by using an antenna that has a narrow vertical beam width.

7.7.3 N = 7/21, 120-deg Capacity

In the $N = 7/21$, 120-deg sectored configuration, the channel capacity is measured by the available voice channels per sector, translated into Erlangs. As a result, a loss of capacity occurs due to trunking inefficiency. This may be evaluated by noting that the number of voice channels per sector is given by $(333 - 21)/21 \approx 14$ in the NES and $(416 - 21)/21 \approx 18$ in the ES. The channel capacity per sector can be evaluated as:

$$\text{NES Channel Capacity} = 14 \text{ voice channels per sector}$$
$$= 8.2 \text{ Erlangs per sector @ 2\% GOS}$$
$$= 3 \times 8.2 = 24.6 \text{ Erlangs/cell @ 2\% GOS} \qquad (7.21)$$

Similarly with the ES, the channel capacity becomes

$$\text{ES Channel Capacity} = 18 \text{ channels per sector}$$
$$= 11.5 \text{ Erlangs per sector @ 2\% GOS}$$
$$= 3 \times 11.5 = 34.5 \text{ Erlangs/cell @ 2\% GOS} \quad (7.22)$$

Table 7.7 summarizes these results for several different GOS values.

7.8 $N = 4/24$ TRICELLULAR PLAN

A tricellular plan is based on a cluster of three identical cells, driven from a single source as shown in Figure 7.13 [7]. Each cell is treated as a logical OMNI, excited from the corner, separated by 120 deg. Unlike the 120-deg sectorization scheme, this scheme enjoys trunking efficiency, reduced hardware and software complexities, reduced MTX/MTSO messaging, and reduced cost.

Table 7.7
$N = 7/21$, 120-Deg Channel Capacity per Cell

Number of Channels per Sector		Channel Capacity per Cell in Erlangs			
		GOS = 1%	GOS = 2%	GOS = 3%	GOS = 5%
NES	14	3 × 7.35 = 22	3 × 8.2 = 24.6	3 × 8.8 = 26.4	3 × 9.73 = 29.19
ES	18	3 × 10.4 = 31.2	3 × 11.5 = 34.5	3 × 12.2 = 36.6	3 × 13.4 = 40.2

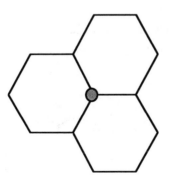

Figure 7.13 A tricellular plan having three logical OMNI cells, excited from the corner, by a single source.

To exploit these benefits, this section presents a method of alternate channel assignment within the N = 4 tricellular plan. The proposed plan is based on dividing the available channels into 12 frequency groups, distributed evenly among four tricells. Frequency assignment is based on an odd/even cyclic distribution of frequency groups, which eliminates adjacent channels and ensures better adjacent channel C/I performance throughout the network. The frequency chart is given in Tables 7.8(a) and (b).

7.8.1 Alternate Channel Assignment

The N = 4, tricellular plan is based on a cluster of four tricells for a total of 12 logical cells. Channel assignment is based on a 4 × 3 array of 12 or multiples of 12 frequency groups, distributed alternately among 12 logical cells (sectors), as shown in Figure 7.14. This arrangement eliminates adjacent channels from adjacent sites, thus reducing adjacent channel interference.

Because there are 12 logical cells and 24 frequency groups (Table 7.8), channel grouping can be formed as n + 12 where n is the number of the logical cell:

Logical cell 1:	Frequency groups:	(1, 13)	
Logical cell 2:	Frequency groups:	(2, 14)	
Logical cell 3:	Frequency groups:	(3, 15)	
Logical cell 4:	Frequency groups:	(4, 16)	
Logical cell 5:	Frequency groups:	(5, 17)	
Logical cell 6:	Frequency groups:	(6, 18)	
Logical cell 7:	Frequency groups:	(7, 19)	
Logical cell 8:	Frequency groups:	(8, 20)	
Logical cell 9:	Frequency groups:	(9, 21)	
Logical cell 10:	Frequency groups:	(10, 22)	
Logical cell 11:	Frequency groups:	(11, 23)	
Logical cell 12:	Frequency groups:	(12, 24)	(7.23)

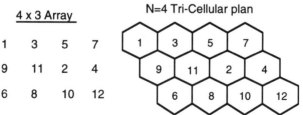

Figure 7.14 The N = 4 tricellular plan having 12 logical cells (sectors) per cluster.

Table 7.8(a)
N = 4/24 Band A Frequency Chart

1	2	3	4	5	6	7	8	9	10	11	12	13	14	15	16	17	18	19	20	21	22	23	24
25	26	27	28	29	30	31	32	33	34	35	36	37	38	39	40	41	42	43	44	45	46	47	48
49	50	51	52	53	54	55	56	57	58	59	60	61	62	63	64	65	66	67	68	69	70	71	72
73	74	75	76	77	78	79	80	81	82	83	84	85	86	87	88	89	90	91	92	93	94	95	96
97	98	99	100	101	102	103	104	105	106	107	108	109	110	111	112	113	114	115	116	117	118	119	120
121	122	123	124	125	126	127	128	129	130	131	132	133	134	135	136	137	138	139	140	141	142	143	144
145	146	147	148	149	150	151	152	153	154	155	156	157	158	159	160	161	162	163	164	165	166	167	168
169	170	171	172	173	174	175	176	177	178	179	180	181	182	183	184	185	186	187	188	189	190	191	192
193	194	195	196	197	198	199	200	201	202	203	204	205	206	207	208	209	210	211	212	213	214	215	216
217	218	219	220	221	222	223	224	225	226	227	228	229	230	231	232	233	234	235	236	237	238	239	240
241	242	243	244	245	246	247	248	249	250	251	252	253	254	255	256	257	258	259	260	261	262	263	264
265	266	267	268	269	270	271	272	273	274	275	276	277	278	279	280	281	282	283	284	285	286	287	288
289	290	291	292	293	294	295	296	297	298	299	300	301	302	303	304	305	306	307	308	309	310	311	312
313	314	315	316	317	318	319	320	321	322	323	324	325	326	327	328	329	330	331	332	333			
									667	668	669	670	671	672	673	674	675	676	677	678	679	680	681
682	683	684	685	686	687	688	689	690	691	692	693	694	695	696	697	698	699	700	701	702	703	704	705
706	707	708	709	710	711	712	713	714	715	716													
															991	992	993	994	995	996	997	998	999
1000	1001	1002	1003	1004	1005	1006	1007	1008	1009	1010	1011	1012	1013	1014	1015	1016	1017	1018	1019	1020	1021	1022	1023

Table 7.8(b)
N = 4/24 Band B Frequency Chart

334	335	336	337	#	339	340	341	342	343	344	345	346	347	348	349	350	351	352	353	354	355	356	357
358	359	360	361	362	363	364	365	366	367	368	369	370	371	372	373	374	375	376	377	378	379	380	381
382	383	384	385	386	387	388	389	390	391	392	393	394	395	396	397	398	399	400	401	402	403	404	405
406	407	408	409	410	411	412	413	414	415	416	417	418	419	420	421	422	423	424	425	426	427	428	429
430	431	432	433	434	435	436	437	438	439	440	441	442	443	444	445	446	447	448	449	450	451	452	453
454	455	456	457	458	459	460	461	462	463	464	465	466	467	468	469	470	471	472	473	474	475	476	477
478	479	480	481	482	483	484	485	486	487	488	489	490	491	492	493	494	495	496	497	498	499	500	501
502	503	504	505	506	507	508	509	510	511	512	513	514	515	516	517	518	519	520	521	522	523	524	525
526	527	528	529	530	531	532	533	534	535	536	537	538	539	540	541	542	543	544	545	546	547	548	549
550	551	552	553	554	555	556	557	558	559	560	561	562	563	564	565	566	567	568	569	570	571	572	573
574	575	576	577	578	579	580	581	582	583	584	585	586	587	588	589	590	591	592	593	594	595	596	597
598	599	600	601	602	603	604	605	606	607	608	609	610	611	612	613	614	615	616	617	618	619	620	621
622	623	624	625	626	627	628	629	630	631	632	633	634	635	636	637	638	639	640	641	642	643	644	645
646	647	648	649	650	651	652	653	654	655	656	657	658	659	660	661	662	663	664	665	666			
717	718	719	720	721	722	723	724	725	726	727	728	729	730	731	732	733	734	735	736	737	738	739	740
741	742	743	744	745	746	747	748	749	750	751	752	753	754	755	756	757	758	759	760	761	762	763	764
765	766	767	768	769	770	771	772	773	774	775	776	777	778	779	780	781	782	783	784	785	786	787	788
789	787	788	789	790	791	792	793	794	795	796	797	798	799										

The growth plan is based on repeating the 4×3 array vertically and horizontally as shown in Figure 7.15. The cell plan is also shown in the same figure where the cell sites are conveniently located to map the $N = 7$ cell pattern. This can be accomplished by selecting a tricell at random and assigning channel numbers according to the $N \times M$ array shown in Figure 7.15.

This scheme also demonstrates tricellular mapping through the network as shown in Figure 7.16. There are seven tricells in the first tier and 12 tricells in the second tier. The first tier is the $N = 7$ tricellular plan with channel reuse within the same cluster. The second tier is formed from a cluster of six surrounding $N = 7$ tricellular pattern, taking two tricells per cluster.

It follows that a given $N = 7$ frequency plan can be translated into an $N = 4$ plan simply by channel reassignment, without cell site relocation. However, this process would require combiner retuning since several channels are combined before transmission.

N x M Array

1	3	5	7	1	3	5	7
9	11	2	4	9	11	2	4
6	8	10	12	6	8	10	12
1	3	5	7	1	3	5	7
9	11	2	4	9	11	2	4
6	8	10	12	6	8	10	12

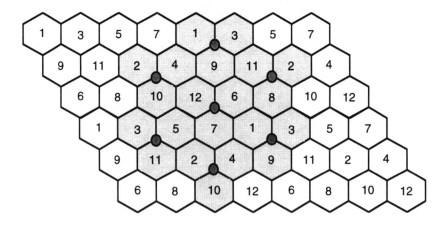

Figure 7.15 The $N = 4$ tricellular growth plan showing $N = 7$ mapping.

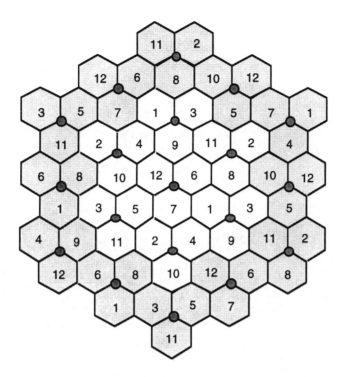

Figure 7.16 The *N* = 4 tricellular growth plan. There are seven tricells in the first tier and 12 tricells in the second tier. The first tier appears as an *N* = 7 cell cluster with channel reuse within the same cluster.

7.8.2 Cyclic Distribution of Channels

In this plan, the channel distribution is based on (*N*, *N* + 4, *N* + 8) where cell number *N* = 1, 2, 3, 4. This is shown in Figure 7.17 where four tricells form a cluster, each cluster having 12 logical cells of 120 deg each. This plan also follows the same channel grouping scheme, described in the previous section, which is formed as *n* + 12 where n is the number of logical cells.

A close inspection of Figure 7.17 will reveal that there is a unique pattern around each tricell. For example, if cell 1 is assumed to be in the center, the surrounding pattern would be 2, 3, 4, 2, 3, 4. Similarly, if cell 2 is in the center, the surrounding pattern would be 1, 3, 4, 1, 3, 4. Likewise, if cell 3 is assumed to be in the center, the surrounding pattern would be 1, 2, 4, 1, 2, 4; and if cell 4 is in the center, the remaining pattern will be 1, 2, 3, 1, 2, 3. This principle can be applied to obtain a growth plan while retaining the adjacent channel isolation throughout a network of any size. This principle is illustrated in Figure 7.18 and a design example

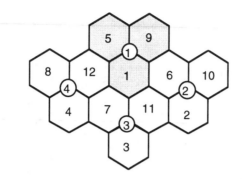

Figure 7.17 The $N = 4$ tricellular plan with cyclic distribution of channels.

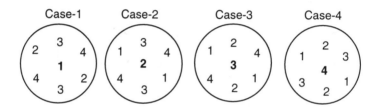

Figure 7.18 Principle of expansion.

is given in Figure 7.19 where cell 1 is assumed to be at the center. Figure 7.19 reveals that there is a unique mapping property between the $N = 4$ and $N = 7$ plans. Therefore, the existing $N = 7$ cell sites can be easily translated into the $N = 4$ plan simply by channel reassignment.

7.8.3 Cochannel Interference in $N = 4$

In $N = 4$, C/I arises from frequency reuse within the same cluster. Thus, referring to Figure 7.19(a), we find that there are three dominant cochannel interferers: one at the back, exactly 180 deg out of phase, and two other interferers from the sides. For example, channel group 3 appears at the back while channel group 11 appears from two sides. The C/I then depends on antenna downtilt and the side-to-side ratio. This may be approximated by the following equation:

$$\frac{C}{I} \approx 10 \log\left[\frac{1}{3}\left(\frac{D}{R}\right)^{\gamma}\right] + \Delta dB(\text{average}) \tag{7.24}$$

where $\Delta dB(\text{average})$ is the average decibel isolation due to antenna downtilt and orientation. With $\gamma \approx 4$, $D/R = 3.732$, and $\Delta dB \geq 6$ dB, we obtain:

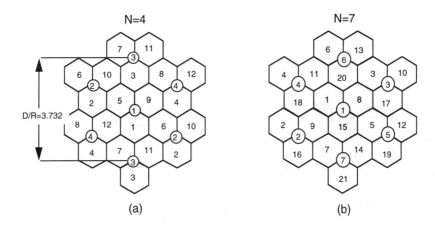

Figure 7.19 The $N = 4$ growth plan mapping into the $N = 7$ pattern. (a) The $N = 4$ plan based on 12 or 24 frequency groups where channels are reused in 21 sectors. (b) The $N = 7$ plan based on 21 frequency groups where channels are not reused within the cluster.

$$C/I \geq 16.8 + 6 = 22.8 \text{ dB} \qquad (7.25)$$

Therefore, the $N = 4$ tricellular plan is a good candidate for urban applications.

7.8.4 $N = 4$ Tricellular Capacity

Channel capacity is determined by means of the available voice channels. Since there are 21 control channels, the available channel capacity can be evaluated as $(333 - 21)/12 \approx 26$ channels per logical cell (sector), $3 \times 26 = 78$ channels per tricell (in the NES). With 2% GOS, this translates into 18.4 Erlangs per logical cell (sector) and $3 \times 18.4 = 55.2$ Erlangs per tricell. Table 7.9 summarizes these results for several grade of services, including expanded spectrum.

Table 7.9
Available Capacity in $N = 4$ Tricellular Plan

| Number of | | Channel Capacity per Cell in Erlangs | | |
Channels per Sector		GOS = 1%	GOS = 2%	GOS = 3%	GOS = 5%
NES	26	$3 \times 17 = 51$	$3 \times 18.4 = 55.2$	$3 \times 11.9 = 57$	$3 \times 21 = 63$
ES	32	$3 \times 22 = 66$	$3 \times 23 = 69$	$3 \times 25 = 75$	$3 \times 26 = 78$

7.9 *N* = 3 TRICELLULAR PLAN

7.9.1 Alternate Channel Assignment

The *N* = 3 tricellular plan is based on a cluster of three tricells for a total of nine logical cells [8]. Channel assignment is based on a 3 × 3 array of nine frequency groups, distributed alternately among nine logical cells as shown in Figure 7.20. Because of alternate channel assignment, this arrangement completely eliminates adjacent channels from adjacent sites, thus reducing adjacent channel interference.

The growth plan is based on repetition of vertical and horizontal patterns in sequence as shown in Figure 7.21. As can be seen, adjacent channel isolation is maintained throughout the network. Note that only vertical and horizontal expansions are possible in this scheme; this is due to the rhombic pattern of the cluster. The *N* = 7 mapping is also evident in Figure 7.21, which can be shown to expand throughout the network.

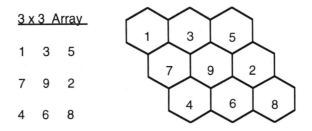

3 x 3 Array

1	3	5
7	9	2
4	6	8

Figure 7.20 The *N* = 3 tricellular plan having nine logical cells (sectors).

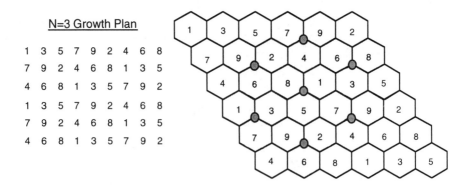

N=3 Growth Plan

1	3	5	7	9	2	4	6	8
7	9	2	4	6	8	1	3	5
4	6	8	1	3	5	7	9	2
1	3	5	7	9	2	4	6	8
7	9	2	4	6	8	1	3	5
4	6	8	1	3	5	7	9	2

Figure 7.21 *N* = 3 tricellular growth plan showing *N* = 7 mapping.

7.9.2 Cyclic Distribution of Channels

In this plan, the channel distribution is based on $(N, N + 3, N + 6)$ where cell number $N = 1, 2, 3$. Using this scheme, we obtain

$$
\begin{aligned}
&\text{Cell 1 frequency groups:} \quad 1, 4, 7 \\
&\text{Cell 2 frequency groups:} \quad 2, 5, 8 \\
&\text{Cell 3 frequency groups:} \quad 3, 6, 9
\end{aligned}
\tag{7.26}
$$

The corresponding $N = 3$ cell cluster is given in Figure 7.22 where the growth plan is already established. For example, if cell 1 is assumed to be in the center, the surrounding pattern would be 2, 3, 2, 3, 2, 3. Similarly, if cell 2 is in the center, the surrounding pattern would be 1, 3, 1, 3, 1, 3. Likewise, if cell 3 is in the center, the surrounding pattern would be 1, 2, 1, 2, 1,2. This is shown in Figure 7.23 where each distribution pattern is free of adjacent channels.

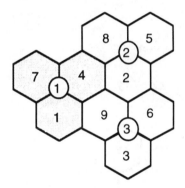

Figure 7.22 The $N = 3$ tricellular plan with cyclic distribution of channels.

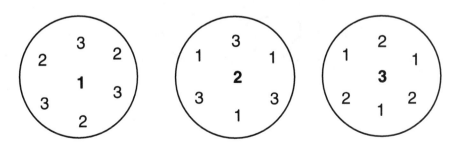

Figure 7.23 Principle of expansion.

Figure 7.24 shows a complete distribution pattern for case 1 where cell 1 is assumed to be in the center. Other distribution patterns can be obtained from the distribution principle according to Figure 7.23.

In Figure 7.24 we also see a mapping similarity between the $N = 3$ and $N = 7$ plans. Therefore, existing $N = 7$ cell sites can be easily translated into an $N = 3$ plan simply by channel reassignment.

7.9.3 N = 3 Cochannel Interference

Referring to Figure 7.24, we find that there are three dominant cochannel interferers from a reuse distance of $D/R = 3$. Thus the cochannel interference can be estimated as:

$$\frac{C}{I} \approx 10 \log\left[\frac{1}{3}\left(\frac{D}{R}\right)^{\gamma}\right] + \Delta dB(\text{average}) \tag{7.27}$$

With $\gamma \approx 4$, $D/R = 3$, and $\Delta dB(\text{average}) \approx 6$ dB, we get

$$C/I \approx 14.3 + 6 = 20.3 \text{ dB} \tag{7.28}$$

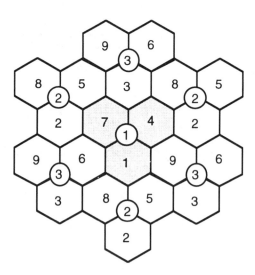

Figure 7.24 The $N = 3$ cyclic channel distribution pattern showing $N = 7$ mapping.

7.9.4 $N = 3$ Tricellular Capacity

Because there are 21 control channels, the available channel capacity may be evaluated as $(333 - 21)/9 \approx 34$ voice channels per logical cell (sector) and $3 \times 34 = 102$ channels per tricell (in the NES). With 2% GOS, this translates into 25.5 Erlangs per logical cell (sector) and $3 \times 25.5 = 76.5$ Erlangs per tricell. Table 7.10 summarizes these results for several different GOS values, including expanded spectrum.

7.10 60-DEG SECTORIZATION

The 60-deg sectorization scheme is achieved by dividing a cell into six sectors of 60 degrees each, as shown in Figure 7.25. Each sector is treated as a logical OMNI cell where directional antennas are used in each sector for a total of six antennas per cell.

Because narrow antenna beam width is used, channels can be repeated more often, thus enhancing the capacity. This configuration is generally used in dense urban environments.

Table 7.10
$N = 3$ Tricellular Capacity

Number of Channels per Sector		Channel Capacity per Cell in Erlangs			
		$GOS = 1\%$	$GOS = 2\%$	$GOS = 3\%$	$GOS = 5\%$
NES	34	$3 \times 23.8 = 71.4$	$3 \times 25.5 = 76.5$	$3 \times 26.8 = 80.4$	$3 \times 28.7 = 86$
ES	44	$3 \times 32.5 = 97.5$	$3 \times 35 = 105$	$3 \times 36 = 108$	$3 \times 38.6 = 115.8$

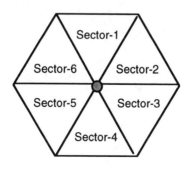

Figure 7.25 Six-sector configuration, 60 deg per sector.

7.10.1 N = 4/24, 60-Deg Sectored Plan

In the $N = 4$, 60-deg frequency reuse plan, the available channels are divided into 24 frequency groups (see Table 7.8), distributed evenly among four 60-deg cell sites, one frequency group per sector. Highly directional antennas are used in each sector, ensuring adequate channel isolation between sectors and better adjacent channel C/I performance throughout the network.

Channel assignment is based on the following sequence: $(N, N + 4, N + 8, N + 12, N + 16$ and $N + 20)$ where N is the cell number ($N = 1, 2, 3$ and 4). Channel grouping scheme is shown in Table 7.11 and the corresponding $N = 4$ cell cluster is shown in Figure 7.26.

Table 7.11
$N = 4/21$, 60-Deg Channel Grouping

Sector: Frequency Group	1 N	2 N + 4	3 N + 8	4 N + 12	5 N + 16	6 N + 20
Cell 1	1	5	9	13	17	21
Cell 2	2	6	10	14	18	22
Cell 3	3	7	11	15	19	23
Cell 4	4	8	12	16	20	24

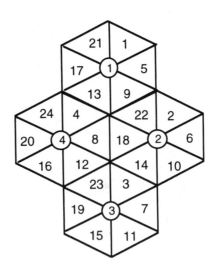

Figure 7.26 The $N = 4/24$ cell planning.

Control channel assignment is based on distributing 21 control channels among 21 sectors; three control channels are reused among the remaining three frequency groups. Alternately, three voice channels are used as control channels at the expense of capacity.

This scheme provides 24×30 kHz = 720-kHz channel isolation within a cluster and 4×30 = 120-kHz isolation between sectors.

7.10.2 $N = 4$, 60-Deg Cochannel Interference

Being $N = 4$, the reuse distance of this plan is

$$D/R = \sqrt{4 \times 3} = 3.46 \tag{7.29}$$

and the cochannel interference (Figure 7.27) is

$$\frac{C}{I} \approx 10 \log\left[\frac{1}{6}(3.46)^4\right] + \Delta dB(\text{average}) \tag{7.30}$$

With a conservative estimate of $\Delta dB \approx 8$ dB, we get

$$C/I \approx 13.8 + 8 = 21.8 \text{ dB} \tag{7.31}$$

which is a good candidate for urban environment.

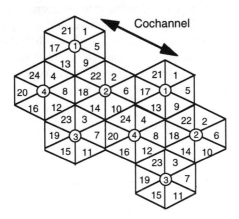

Figure 7.27 Cochannel interference due to frequency reuse

7.10.3 $N = 4/24$, 60-Deg Capacity

Because there are 24 control channels in this scheme, the number of voice channels in this scheme is $333 - 24 = 309$ in the NES and $416 - 24 = 392$ in the ES. The channel capacity per sector can now be evaluated as:

$$\text{NES channel capacity} = (309)/24$$
$$= 12 \text{ channels per sector}$$
$$= 6.61 \text{ Erlangs/sector @ 2\% GOS}$$
$$= 6 \times 6.61 = 39.66 \text{ Erlangs/cell @ 2\% GOS}$$

Similarly with ES, the channel capacity becomes

$$\text{ES channel capacity} = 392/24 \approx 16 \text{ channels per sector}$$
$$= 9.8 \text{ Erlangs/sector @ 2\% GOS}$$
$$= 6 \times 9.8 = 58.8 \text{ Erlangs/cell @ 2\% GOS}$$

Table 7.12 summarizes these results for several different GOS values.

7.11 $N = 3$, 60-DEG SECTORIZATION

The $N = 3$, 60-deg sectorization is based on distributing the available channels among three cells where each cell is divided into six sectors. In this configuration, the available channels are divided into 18 frequency groups, distributed evenly among three 60-deg cell sites. Frequency assignment is based on alternate channel distribution, which ensures adequate channel isolation between sectors and better adjacent channel C/I performance throughout the network. This scheme is also suitable for urban areas for enhancing channel capacity.

Table 7.12
$N = 4/24$, 60-Deg Channel Capacity

Number of Channels/Sector		Channel Capacity in Erlangs/Cell				
		GOS = 1%	GOS = 2%	GOS = 3%	GOS = 4%	GOS = 5%
NES	12	36	39.66	43		48
ES	16	53	58	62		69

7.11.1 $N = 3$, 60-Deg Plan

The $N = 3$ frequency plan is based on dividing the available channels into 18 frequency groups (see Table 7.4). This plan provides $333/18 \approx 18$ channels per group in the NES and $416/18 \approx 23$ channels per group with ES. The cell cluster is based on three cells, six sectors per cell, for a total of 18 sectors per cluster. Channel distribution is based on cyclic distribution of channels, which ensures adequate channel isolation between sectors and better adjacent channel C/I performance throughout the network. This is shown in Figure 7.28 where one frequency group is assigned per sector for a total of $6 \approx 23 = 138$ channels per cell. This translates into 38% channel enhancement over the existing $N = 4$ frequency plan. Note that in TDMA-3, the capacity is further enhanced threefold.

7.11.2 Evaluation of Cochannel Interference

With $N = 3$, the repeat distance is given by

$$d_i/d_c = \sqrt{3N} = 3 \tag{7.32}$$

and the C/I (Figure 7.29) is given by

$$\frac{C}{I} \approx 10 \log\left[\frac{1}{6}(3)^4\right] + \Delta dB(\text{average}) \tag{7.33}$$

With $\Delta dB \approx 8$ dB, we obtain:

$$C/I \approx 11.3 + 8 = 19.3 \text{ dB} \tag{7.34}$$

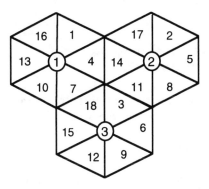

Figure 7.28 The $N = 3$, 60-deg cell plan based on three cells per cluster.

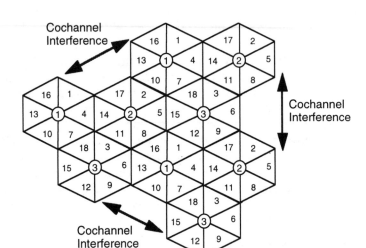

Figure 7.29 The $N = 3$ cochannel interference evaluation scheme.

which is marginal. Therefore antenna directivity, downtilt, etc., are critical in this configuration. Note that the minimum C/I requirement is 17 dB in AMPS.

The channel capacity per sector in this plan is limited to $333/24 \approx 13$ compared to 18 in the $N = 3$ plan (NES) and $416/24 \approx 17$ compared to 23 in the $N = 3$ plan (ES).

7.12 DIRECTIONAL FREQUENCY REUSE

From the preceding sections we conclude that there are two critical parameters in frequency planning:

1. High channel count per cell or sector;
2. Low cochannel interference.

These parameters also oppose each other; that is, high channel count reduces C/I and vice versa. We also realize that the overall C/I performance depends on the sum of all interferers. Therefore, a possible solution to this problem would be to reduce the number of interferers and reuse channels more often, thereby enhancing the capacity while maintaining the C/I performance at an acceptable level [9].

In this section we present a process of directional frequency reuse. The process starts by creating a quantity of tricell groups from three cells. Seven of the tricell groups are then formed into a cellular cluster where the antenna front-to-back ratio is exploited to isolate a repeat frequency. As a result, a frequency can be reused more often, thereby augmenting the cellular capacity.

7.12.1 Directional Frequency Reuse Plan

The proposed directional frequency reuse plan is based on a tricellular platform using a cluster of three identical cells, driven from a single source as shown in Figure 7.30. An antenna array comprised of directional antennas is positioned relatively close to the center of each tricell group. Each antenna radiates into a predetermined sector of the tricell group. Because directional antennas are used in each sector, maximum cochannel isolation is available only in three directions, 120 deg apart. Therefore for maximum cochannel isolation, frequencies should be reused in these three directions only. For example, channel group A and channel group B may be reused alternately only at 0 deg without repeating in any other directions. Similarly, channel group C and channel group D may be reused alternately at 120 deg without repeating in any other directions. Likewise, channel group E and channel group F may be reused alternately at 240 deg without repeating in any other directions.

7.12.2 Directional Frequency Reuse Growth Plan

Figure 7.31 illustrates the principle of the growth plan within a seven tricellular platform. There are three directions, 120 deg apart. In each direction there exist three parallel layers, a, b, and c, as described in the following equations:

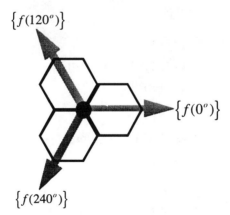

Figure 7.30 Illustration of directional frequency reuse where a frequency group is permitted to be reused in one direction only.

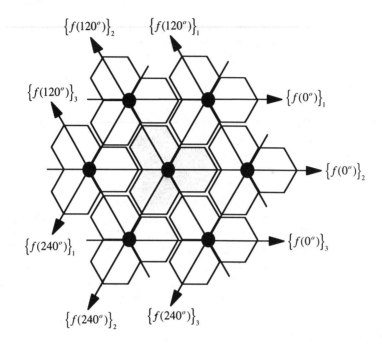

Figure 7.31 A tricellular platform illustrating directional frequency reuse.

$$\{ f_a(0°)\}$$
$$\{ f_b(0°)\} \qquad \text{at 0 deg} \qquad (7.35)$$
$$\{ f_c(0°)\}$$

$$\{ f_a(120°)\}$$
$$\{ f_b(120°)\} \qquad \text{at 120 deg} \qquad (7.36)$$
$$\{ f_c(120°)\}$$

$$\{ f_a(240°)\}$$
$$\{ f_b(240°)\} \qquad \text{at 240 deg} \qquad (7.37)$$
$$\{ f_c(240°)\}$$

There are a total of nine layers, three in each direction where a set of frequencies is permitted to be reused in a particular direction only. Therefore, the directional frequency reuse plan is based on dividing up the available {395 voice channels into predetermined groups. These groups are then directionalized and distributed according to the proposed scheme.

Example

As an example we consider a directional reuse for $N = 4$ having 12 frequency groups. Three pairs of alternate frequency groups are directionalized and reused according to the following principle:

$$\{f_a(0°)\} = 1, 3, 1, \ldots$$
$$\text{at } 0 \text{ deg} \qquad (7.38)$$
$$\{f_{b,c}(0°)\} = 2, 4, 2 \ldots$$

$$\{f_a(120°)\} = 5, 7, 5, \ldots$$
$$\text{at } 120 \text{ deg} \qquad (7.39)$$
$$\{f_{b,c}(120°)\} = 6, 8, 6 \ldots$$

$$\{f_a(240°)\} = 9, 11, 9, \ldots$$
$$\text{at } 240 \text{ deg} \qquad (7.40)$$
$$\{f_{b,c}(240°)\} = 10, 12, 10 \ldots$$

The corresponding frequency reuse plan is shown in Figure 7.32 where each frequency group is positioned along the direction of the antenna which is consistent throughout the service area. The total number of interferers is therefore three, and among them only one is the dominant interferer, which is at the backlobe of the antenna, and the other two run parallel, exactly 90 deg out of phase, providing sufficient isolation compared to the one in the backlobe.

C/I due to the dominant interferer is given by:

$$\frac{C}{I} \approx 10 \, \log[(3.732)^4] = 22.87 \text{ dB} \qquad (7.41)$$

The channel capacity of this plan is $395/12 = 33$ channels per sector and $33 \times 3 = 99$ channels per cell. At 2% GOS, this translates into 24.6 Erlangs per sector and $24.6 \times 3 = 73.8$ Erlangs per cell.

7.13 SAT PLANNING

The supervisory audio tone (SAT) is a set of three frequencies: 5970, 6000, and 6030 Hz. These tones are generated in the base station and transmitted to mobile unit over the voice channels. Upon receiving the SAT tone, the mobile returns it to the base station as an acknowledgment. The purpose of the SAT tone is to distinguish between cochannels. This means that the *same SAT cannot be reused in a cochannel site.*

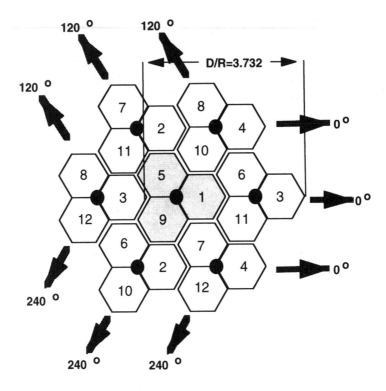

Figure 7.32 The $N = 4$ directional frequency reuse plan.

Since there are only three SAT tones:

$$S0 = 5970 \text{ Hz}$$
$$S1 = 6000 \text{ Hz}$$
$$S2 = 6030 \text{ Hz}$$

their distribution pattern is restricted to two versions only, as described in the next two sections.

7.13.1 Version 1: Alternate SAT Distribution Within a Cluster

In the $N = 7$ frequency plan, the available channels are evenly distributed among seven cells without repetition. Therefore, if we assign SAT-0 in the center cell, the surrounding pattern would be S1, S2, S1, S2, S1, S2. Similarly, if S1 is assigned in the center cell, the surrounding pattern would be S0, S2, S0, S2, S0, S2. Likewise, if S2 is assigned in the center cell, the surrounding pattern would be S0, S1, S0, S1, S0, S1. This principle is illustrated in Figure 7.33 and the growth plan is given in Figure 7.34.

7.13.2 Version 2: SAT Reuse Within a Cluster

It is also possible to use the same SAT within the same cluster as shown in Figure 7.35. A close inspection will reveal that this configuration does not allow voice channel reuse within the same cluster since the SATs are identical. The growth plan is shown in Figure 7.36.

7.14 DIGITAL COLOR CODE PLANNING

Digital color code (DCC) is a 2-bit code used to identify control channels. These codes are also generated in the base station and transmitted to the mobile over the

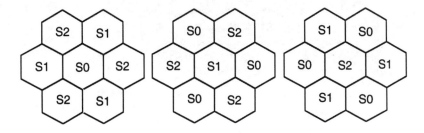

Figure 7.33 Principle of alternate SAT distribution within a cluster.

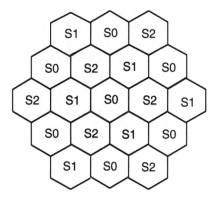

Figure 7.34 SAT growth plan.

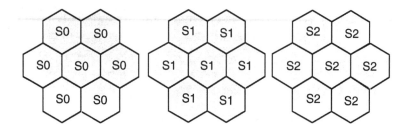

Figure 7.35 Principle of SAT reuse within a cluster.

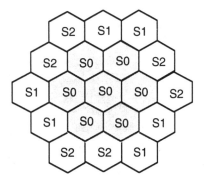

Figure 7.36 SAT Growth plan.

control channel. The purpose of DCC is to differentiate between control channels in cochannel sites. This means that the *same DCC cannot be reused in a cochannel site.*

Because there are only four DCCs (D0 = 00, D1 = 01, D2 = 10, D3 = 11), their distribution pattern is restricted. A commonly used distribution pattern is described in Figure 7.37. The principle is as follows: If D0 is assigned in the center cell, the surrounding pattern would be D1, D2, D3, D1, D2, D3. This is the first tier of cells. The second tier contains 12 cells whose distribution can be obtained by visualizing each cell as the center of a seven-cell cluster. Then if D1 is assumed to be the center of a seven-cell cluster, the surrounding pattern can be readily obtained from the distribution pattern established in the first tier, which is D0, D2, D3, D0, D2, D3. Likewise, if D2 is assumed to be the center cell, the surrounding pattern would be D0, D1, D3, D0, D1, D3. On the other hand, if D3 is assumed to be the center cell, the surrounding pattern would be D0, D1, D2, D0, D1, D2. A complete growth plan is shown in Figure 7.38.

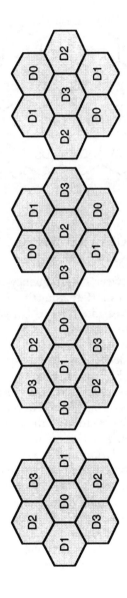

Figure 7.37 Principle of DCC planning.

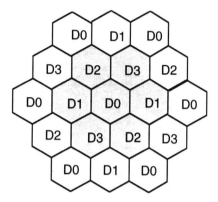

Figure 7.38 DCC growth plan.

7.15 A CLASSICAL METHOD OF EVALUATING COCHANNEL AND ADJACENT CHANNEL INTERFERENCE

The mechanism of interference between two frequency-modulated (FM) signals was first examined by Corrington [10] in 1946, shortly after the development of FM. He demonstrated that the effect of interference between two modulated carrier frequencies is to produce cross-modulation between two signals. The beat note produced by this interference consists of a series of sharp peaks and dips of noise, which are superimposed on the desired audio output. As a result the channel is subject to an outage, link failure, FM capture, etc.

In this section, the mechanism of interference between two FSK-modulated carriers is examined [11] by means of the method developed by Corrington for FM. It is shown that, in digital FM, the deviation in the instantaneous frequency due to interference causes a corresponding deviation in the eye opening [12], thereby resulting in a corresponding degradation in the signal-to-noise ratio (SNR). Here, the term *eye opening* refers to the base-band binary signal, which looks like a human eye when displayed on an oscilloscope. The eye opening is an indication of the signal quality: The height of this opening decreases as noise, interference, etc., are introduced into the channel. We show that in cochannel channel interference, the eye opening is independent of the modulation index, whereas in adjacent channel interference the eye opening depends on the modulation index as well as on adjacent channel separation. These findings are presented in this section.

7.15.1 Cochannel Interference

In cochannel interference, the effect of interference between two modulated carrier frequencies is to produce cross-modulation between two signals. The beat note produced by this interference consists of a series of sharp peaks and valleys of noise, which are superimposed on the desired audio output. The envelope of the instantaneous frequency of the desired signal has two components given by [10]:

$$F_i(\text{peaks}) = \frac{\Delta f}{1+x}\cos(\Omega 1 t) + \frac{x \cdot F_d}{1+x}\cos(\Omega 2 t) \tag{7.42}$$

$$F_i(\text{dips}) = \frac{\Delta f}{1-x}\cos(\Omega 1 t) - \frac{x \cdot F_d}{1-x}\cos(\Omega 2 t) \tag{7.43}$$

where

F_i	=	instantaneous frequency
Δf	=	frequency deviation of signal 1 due to signal 2
F_d	=	frequency separation between signal 1 and signal 2
x	=	I/C interference-to-carrier ratio
$\Omega 1$	=	Modulating frequency of signal 1
$\Omega 2$	=	Modulating frequency of signal 2.

In the absence of interference $x = 0$ and the envelope of the instantaneous frequency becomes

$$F_i(\text{ideal}) = \Delta f \cos(\Omega 1 t) \tag{7.44}$$

With increased interference, deviation in the envelope also increases. Thus, by assuming that the fractional deviation of the instantaneous frequency causes a fractional deviation in the eye opening (H), which in turn depends on the ratio of envelopes of the peak deviation to the ideal signal, we write

$$H = \frac{F_i(\text{peaks})}{F_i(\text{ideal})} \tag{7.45}$$

where H is the eye opening.
From (7.42), (7.44), and (7.45) we obtain:

$$H = \frac{1}{1+x} + \frac{F_d}{\Delta f} \times \frac{x}{1+x} \times \frac{\cos(\Omega 2 t)}{\cos(\Omega 1 t)} \tag{7.46}$$

In common channel interference $F_d = 0$, and (7.46) reduces to

$$H(\text{common channel}) = \frac{1}{1+x} \tag{7.47}$$

where $x = (I/C)$, measured in voltage. It also indicates that, in common channel interference, the SNR is independent of the modulation index but depends on the interference-to-signal ratio. The degradation of SNR is given by

$$d(\text{SNR}) = 20 \log(H) \qquad (7.48)$$

Thus by knowing the eye opening, the degradation in SNR and hence the bit error rate (BER) can be easily computed. Table 7.13 shows the degradation of SNR due to several values of I/C. Table 7.13 indicates that a 3-dB degradation of SNR (71% eye opening) requires a C/I ratio of 4, which translates into a C/I of 12 dB. Note that in FM, the C/I ratio of 3 (9.5 dB) is generally accepted, after which the amount of distortion in the audio increases rapidly.

7.15.2 Adjacent Channel Interference

In adjacent channel interference $F_d \neq 0$. The eye opening appears as [see (7.46)]:

$$H = \frac{1}{1 + x} + \frac{F_d}{\Delta f} \times \frac{x}{1 + x} \times \frac{\cos(\Omega 2t)}{\cos(\Omega 1t)} \qquad (7.49)$$

Maximum interference occurs when the carriers approach each other, due to FSK modulation. This implies that $\cos(\Omega 2t) = 1$ and $\cos(\Omega 1t) = -1$ or vice versa. In either case the ratio is -1. Therefore, in worst case interference, the ratio $\cos(\Omega 2t)/\cos(\Omega 1t) = -1$ and (7.49) reduces to

Table 7.13
Cochannel Interference

x (= I/C)	Eye Opening (H)	Degradation of SNR (dB)
0.0	1	0.0
0.1	0.91	−0.8
0.2	0.833	−1.6
0.3	0.77	−2.3
0.4	0.714	−2.9
0.5	0.666	−3.5
0.6	0.625	−4.1
0.7	0.59	−4.6
0.8	0.555	−5.1
0.9	0.52	−5.56
1.0	0.5	−6.0

$$H = \frac{1}{1 + x} - \frac{F_d}{\Delta f} \times \frac{x}{1 + x} \qquad \text{(adjacent channel)} \qquad (7.50)$$

which indicates that, in adjacent channel interference, the eye opening is impaired by channel separation as well as by the modulation index. However, the analysis can be further simplified by noting that worst case interference occurs when the frequency deviation, $\Delta f = F_d/2$ on either side of the center frequency, which means that the total allowed deviation is Fd. Therefore the ratio $F_d/\Delta f = 1$. The eye opening in the worst case adjacent channel interference then becomes

$$H = \frac{1 - x}{1 + x}$$

With $x = I/C$, we write

$$H = \frac{1 - I/C}{1 + I/C} \qquad (7.51)$$

The degradation of SNR is given in (7.48). Table 7.14 shows the dependence of eye opening on I/C. Table 7.14 indicates that the eye opening is highly sensitive to adjacent channel interference. This is due to the occurrence of beat frequencies between the two modulated carriers. The 3-dB degradation of SNR corresponds to an I/C of 0.15, which translates into a C/I of 16 dB.

Table 7.14
Adjacent Channel Interferences

$x(= I/C)$	Eye Opening (H)	Degradation of SNR (dB)
0.0	1	0.0
0.1	0.82	−1.74
0.2	0.666	−3.52
0.3	0.54	−5.37
0.4	0.43	−7.36
0.5	0.333	−9.54
0.6	0.25	−12
0.7	0.176	−15
0.8	0.111	−19
0.9	0.05	−25.6
1.0	0.0	−∞

7.16 CONCLUSIONS

We have presented several frequency planning methods along with C/I performance and channel capacity. The most popular OMNI frequency plan is the $N = 7$ plan, which also exhibits adjacent channel interference. We have shown that the $N = 9$ OMNI frequency plan outperforms the $N = 7$ plan with respect to C/I performance but suffers from reduced channel capacity.

Various sectored plans were then examined along with the tricellular plan. The $N = 4$ tricellular plan appears to be a better compromise between the channel capacity and C/I performance. It is further shown that a given $N = 7$ plan can be easily translated into an $N = 4$ or a $N = 3$ tricellular plan by channel reassignment, requiring no cell site relocation.

We also conclude that a compromise between channel capacity and C/I is critical in frequency planing. A method for a directional reuse plan was then presented to acquire both capacity and performance.

Finally, the mechanism of interference between two modulated carriers was examined and C/I prediction equations presented. We have shown that in common channel interference, the C/I performance is independent of the modulation index, while in adjacent channel interference the C/I performance depends on the modulation index as well as on adjacent channel separation.

In conclusion, frequency planning is an important engineering task in cellular communication systems. It not only determines the system performance and capacity but also governs the overall system growth and revenue.

References

[1] IS-54, "Dual-Mode Mobile Station-Base Station Compatibility Standard," PN-2215, Electronic Industries Association Engineering Department, December 1989.

[2] Lee, William C. Y., *Mobile Cellular Telecommunications Systems*, New York: McGraw-Hill Book Company, 1989.

[3] Mehrotra, Asha, *Cellular Radio, Analog and Digital Systems*, Norwood, MA: Artech House, 1994.

[4] Mehrotra, Asha, *Cellular Radio Performance Engineering*, Norwood, MA: Artech House, 1994.

[5] Benner, F., "Method for Assigning Telecommunications Channels in a Cellular Telephone System," U.S. Patent 5,111,534, May 5, 1992.

[6] Faruque, Saleh, "The $N = 9$ Frequency Plan: A Modified Technique to Enhance C/I Performance and Capacity," *Proc. IEEE 2nd Int. Conf. on Universal Personal Communications*, Ottawa, Ontario, Canada, 1993, pp. 718–722.

[7] Faruque, Saleh, "N = 4 Tricellular Plan with Alternate Channel Assignment," MILCOM-95.

[8] Faruque, Saleh, "A Non-Interfering Frequency Plan for Cellular Communication Systems," U.S. Patent Pending.

[9] Faruque, Saleh, "Directional Frequency Assignment in a Cellular Radio System," U.S. Patent Pending.

[10] Corrington, M. S., "Frequency Modulation Distortion Caused by Common and Adjacent Channel Interference," *RCA Review*, Vol. 7, December 1946, pp. 522–560.

[11] Faruque, Saleh, "A Classical Method of Evaluating Cochannel and Adjacent Channel Interference in Digital FM," MILCOM-95.

[12] Smith, David R., *Digital Transmission Systems*, New York: Van Nostrand Reinhold Company, 1985.

CHAPTER 8
▼▼▼

CELL SITE ENGINEERING

8.1 INTRODUCTION

A dual-mode cellular base station (Figure 8.1), supporting both AMPS and TDMA, is essentially a multiple-access radio communication center where several radio-frequency (RF) channels are combined to form a channel group, which is then assigned to a cell or a sector.

The cell site is then connected to the mobile switching center through a cross point switch via T_1 links. The cross-point switch also converts T_1 data from serial to parallel and parallel to serial format. Each channel is assigned to a base station radio where radio port assignment to a channel is performed during cell site engineering. Once the cell site is configured, the radio port assignment cannot be changed dynamically. An array of such base stations has the capacity of serving tens of thousands of subscribers in a major metropolitan area where both AMPS and TDMA services may coexist in the same base station.

As shown in Figure 8.1, the major components of a cellular base station are:

- Cellular radio;
- Combiners;
- Remote multicoupler (RMC);
- Duplexers;

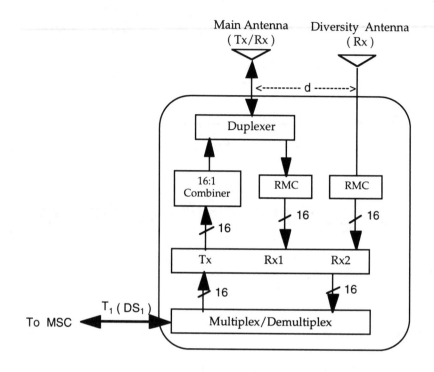

Figure 8.1 Cellular base station comprising several radios and supporting equipment (Tx = transmitter, Rx₁ = main receiver, Rx₂ = diversity receiver).

- Antennas;
- Multiplexers.

These components are housed in a specially designed, temperature-controlled equipment shelter. Cell site engineering, therefore, is a step-by-step process of base station design, system integration, and deployment involving:

- Business planning;
- Site selection and coverage prediction;
- Radio link design (link budget);
- Traffic engineering and base station provision;
- Base station design and system integration;
- Cell planning;
- Channel assignment;
- Antenna engineering.

In short, it combines science, engineering, and art, where a good compromise among all three is the key to the successful implementation and continued healthy operation of cellular base stations.

In this concluding chapter we present a concise engineering process involved in various stages of the design and deployment of dual-mode AMPS-TDMA cellular base stations.

8.2 INITIAL PLANNING AND DESIGN CONSIDERATIONS

8.2.1 Site Selection for Dual Mode AMPS-TDMA

Site selection and coverage prediction is an iterative process of selecting a cell site location and predicting the RF footprint (RF coverage). Initially, the site selection process involves several site visits and customer interface to prescreen sites by means of a spectrum analyzer as shown in Figure 8.2. The objective of scanning the neighborhood is to detect:

- Competing cellular operators;
- Competing voice and control channel signal strengths;
- Other potential interferers such as intermod products.

By knowing the frequencies, the corresponding channel number can be computed from the following equations [1]:

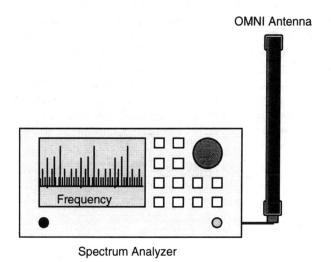

Figure 8.2 Spectrum analyzer used as a scanning receiver.

$$\text{Transmitting frequency} = (0.03N + 870) \text{ MHz} \qquad \text{(NES)}$$
$$= 0.03(N - 1023) + 870 \text{ MHz} \quad \text{(ES)} \qquad (8.1)$$

$$\text{Receiving frequency} = 0.03N + 825 \text{ MHz} \qquad \text{(NES)}$$
$$= 0.03(N - 1023) + 825 \text{ MHz} \quad \text{(ES)} \qquad (8.2)$$

where N is the channel number ($N = 1, 2, \ldots, 1023$). A set of channels can be identified as cellular channels while the rest are intermod products. This information is useful not only in screening cell sites but also in achieving optimum frequency planning. A mutual understanding with the competing operators may have to be reached in terms of spectrum control so that both frequency bands (Bands A and B) can coexist without interfering with each other. A set of candidate sites is then enlisted for coverage prediction, traffic analysis, and cell site provisioning

Antenna height and location play an important role in determining the RF footprint in a given propagation environment. Rooftop antenna exhibits high pathloss slope while an antenna placed below the rooftop exhibits the waveguide effect in urban and dense urban environments. Therefore, it is also desirable to note the description of each site. Moreover, all candidate sites must have good access for installation and services.

8.2.2 Tolerance of Cell Site Location

Often, it is not possible to install a cell site in the desired location due to physical restrictions, so the cell site has to be relocated, preferably in a nearby location. As a result, the D/R ratio will change, affecting the carrier-to-interference (C/I) ratio. In this section we examine the degradation of C/I due to cell site relocation and determine the maximum allowable relocation distance for which C/I $\geq$ 18 dB.

Consider a pair of cochannel sites having a reuse distance D as shown in Figure 8.3. Because of geography and physical restrictions, both cell sites have to be relocated. Let's assume that both cell sites approach each other by ΔD. The new reuse distance is therefore $D - 2\Delta D$. The corresponding C/I then becomes

$$\frac{C}{I} = 10 \log\left[\frac{1}{j}\left(\frac{D - 2\Delta D}{R}\right)^{\gamma}\right] \qquad \text{for } 0 < \alpha < 1 \qquad (8.3)$$

With $\dfrac{D}{R} = \sqrt{3N}$ and $\alpha = \Delta D/D$, we obtain

$$\frac{C}{I} = 10 \log\left[\frac{1}{j}\{\sqrt{3N}(1 - 2\alpha)\}^{\gamma}\right] \qquad \text{for } 0 < \alpha < 1 \qquad (8.4)$$

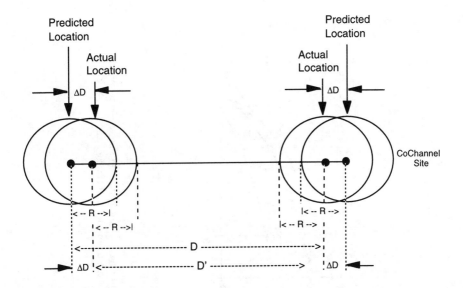

Figure 8.3 Cell site location tolerance.

where j = number of cochannel interferers, N = frequency reuse plan, γ = path-loss slope and $0 < \alpha < 1$. Equation (8.4) is plotted in Figure 8.4 for two different frequency reuse plans, $N = 4$ and $N = 7$. We see that the $N = 4$ frequency plan has zero tolerance on cell site relocation. Moreover, the $N = 4$ OMNI does not meet the C/I requirement of 18 dB. Additional measures such as sectoring are needed to acquire the additional margin. On the other hand, the $N = 7$ OMNI meets the 18-dB C/I requirement in dense urban environment ($\gamma = 4$) even with 10% variation of the cell site location. However, it has zero tolerance in urban ($\gamma = 3.5$) and suburban ($\gamma = 2.5$) environments.

It follows that a number of engineering considerations are involved at this stage before going further. These are (1) C/I versus capacity, (2) the chosen frequency reuse plan, (3) OMNI versus sectorization, and (4) the propagation environment.

8.2.3 C/I Versus Capacity

The cellular communication system is a multiple-access network where several noninterfering channels are combined to form a channel group, which is then assigned to a cell or a sector. These channel groups are reused at regular distance intervals, thus enhancing the channel capacity. However, this is accomplished at the expense of C/I performance. Therefore, a compromise has to be attained so that the target capacity is acquired without jeopardizing C/I performance. As an illustration let's consider the familiar cochannel interference prediction equation as follows:

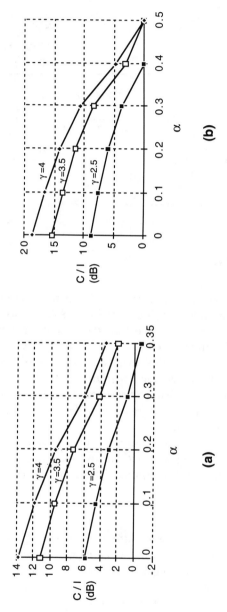

Figure 8.4 C/I as a function of α in different propagation medium: (a) $N = 4$ OMNI frequency reuse plan, requiring sectoring for an additional C/I margin; (b) $N = 7$ OMNI frequency reuse plan, requiring sectoring for suburban and rural environments.

$$\frac{C}{I} = 10 \, \log\left[\frac{1}{j}\{\sqrt{3N}\}^{\gamma}\right] + \Delta dB|_{sector} \qquad (8.5)$$

where N is the frequency reuse plan such as 3, 4, 7, 9, 12, etc.; j is the total number of cochannel interferers; γ is the path-loss slope; and ΔdB is an additional margin due to sectorization, shadowing, etc. Because frequency planning is a means to distribute 395 voice channels equally among a cluster of cells, the channel capacity per cell will depend on the frequency reuse plan. Thus, the channel capacity can be determined as

$$\text{Number of voice channels per cell} = \frac{395}{N} \quad \text{for } N = 3, 4, 7, 9, 12 \quad (8.6)$$

With $\gamma = 4$, $j = 6$, grade of service (GOS) = 2%, $\Delta dB = 0$, and channel capacity translated into Erlangs, we obtain the relationship between C/I and Erlang capacity as a function of frequency reuse plan (see Table 8.1; it is also plotted in Figure 8.5).

Table 8.1
C/I and Erlang Capacity As a Function of Frequency Reuse Plan (Path-Loss Slope = 4, Six Interferers, GOS = 2%)

Number of VCH/Cell	N	C/I (dB)	Erlang Capacity
131	3	11.3	119
98	4	13.8	86
56	7	18.6	46
43	9	20.8	34
32	12	23.3	24

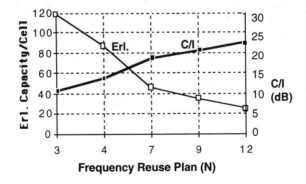

Figure 8.5 C/I and Erlang capacity as a function of frequency reuse plan.

From the preceding analysis we see that a frequent reuse of channels enhances capacity at the expense of C/I performance. Therefore, a compromise is needed so that the capacity is enhanced and the C/I objective is also met. Note that an additional C/I margin can be acquired by means of sectorization and the utilization of buildings as a shield. The bottom line is that the designer has the crucial task of melding theory and practice. That is to say, the following items are the key determining factors for successful implementation of a cell site that not only performs well but also generates revenue:

- Analytical background;
- Experience;
- Site visit;
- Best judgment.

8.2.4 Cell Planning and Frequency Planning Considerations

The next step of the process is to choose the frequency plan best suited for the proposed service. There are several frequency plans (e.g., the N = 9, 7, 4, or 3 frequency reuse plan) so a choice has to be made as to which frequency plan is appropriate for the propagation environment that also meets the traffic requirements.

Here, our main objective is to examine the available frequency plans as a function of propagation environment by means of (8.5), which is plotted for N = 4, 7, and 9 in Figure 8.6. We see that an OMNI frequency plan that can be

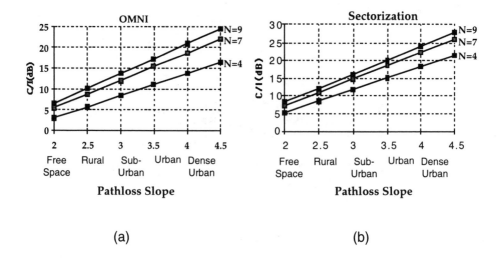

(a)   (b)

Figure 8.6 C/I as a function of path-loss slope: (a) OMNI and (b) sectored, worst case.

used in urban and dense urban environments ($\gamma \approx 4$)) is $N \geq 7$. For suburban and rural environments, $\gamma < 3.5$, these cell sites would have to be sectored for an additional margin so that C/I ≥ 18 dB. Alternately, a frequency reuse plan much greater than $N = 7$ would have to be used at the expense of capacity; this might be an acceptable solution because the traffic density outside the city is generally less. On the other hand, $N < 7$ frequency reuse plans can be used in urban and dense urban environments if they are sectored and buildings are used as shields in order to satisfy the following equation:

$$\frac{C}{I} = 10 \log\left[\frac{1}{j}\{\sqrt{3N}\}^{\gamma}\right] + \Delta dB|_{sector} + \Delta dB|_{building} \tag{8.7}$$

where

$$10 \log\left[\frac{1}{j}\{\sqrt{3N}\}^{\gamma}\right] = C/I \text{ due to OMNI}$$

$\Delta dB|_{sector}$ = Additional C/I margin due to sectorization

$\Delta dB|_{building}$ = Additional C/I margin due to buildings, used as shields

Several conclusions can be drawn from the preceding discussions:

- $N \geq 7$ OMNI sites are good candidates for dense urban and urban environments ($\gamma \geq 4$).
- $N \geq 7$ sectored sites are good candidates for dense urban ($\gamma > 4$), urban ($\gamma = 4$), and suburban ($\gamma \approx 3$ to 3.5) environments.
- $N < 7$ OMNI sites are not good candidates for outdoor application.
- $N < 7$ sectored sites are good candidates for dense urban environments.
- $N < 7$ sectored sites are not good candidates for suburban and rural environments.

Note also that the $N = 7$ OMNI plan provides more capacity than the $N = 7$ sectored plan due to trunking efficiency (see Chapter 6). On the other hand, $N = 7$ sectored sites mitigate multipath components. Therefore if capacity is not an issue, then $N = 7$ OMNI plan should be used in dense urban and urban environments. Otherwise $N = 7$ sectorization should be the cell plan of choice. Furthermore, if the capacity is a primary concern, then $N = 4$ tricellular plan should be chosen for dense urban and urban environments and the $N \geq 7$ sectored plan should be chosen for suburban and rural environments. These scenarios are depicted in Figure 8.7.

We also note that antenna location plays an important role in achieving an additional C/I margin, needed for $N < 7$ frequency plans. Figure 8.8 shows a deployment scenario in a downtown core where each base station antenna is shielded by

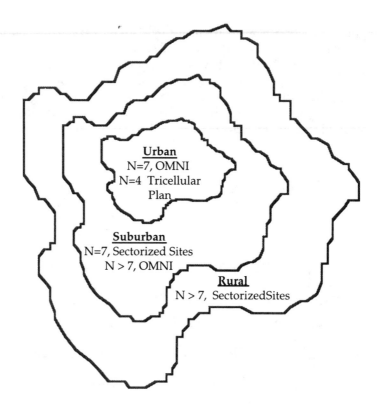

Figure 8.7 Possible cell planning and deployment guide for urban, suburban, and rural environments.

buildings. The typical margin that can be achieved by this method is >10 dB, which is adequate for $N = 3$ and $N = 4$ tricellular plans.

8.2.5 C/I Versus Antenna Height

We have demonstrated that C/I is a function of frequency reuse factor N and propagation path-loss slope γ. This is analogous to C/I as a function of the propagation environment, which in turn depends on antenna height. To examine this phenomenon, let's consider the conventional C/I relationship once again as shown here:

$$\frac{C}{I} = 10 \, \log\left[\frac{1}{j}\{\sqrt{3N}\}^{\gamma}\right] \tag{8.8}$$

where γ is a function of antenna height determined by the appropriate propagation model (see Chapter 5):

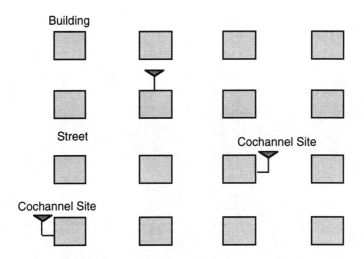

Figure 8.8 Antenna placement where buildings are used as shields.

$$\gamma = \left\{ \frac{44.9 - 6.55 \ \log(h_b)}{10} \right\} \qquad \text{Okumura-Hata model} \qquad (8.9)$$

$$\gamma = \left\{ \frac{20 + k_d}{10} \right\} \qquad \text{Walfisch-Ikegami} \qquad (8.10)$$

where

k_d = $18 - 15 \ (\Delta h_b / h_{\text{roof}})$
h_b =   base station antenna height
$\dfrac{\Delta h_b}{h_{\text{roof}}}$ =   ratio of rooftop antenna height to the building height.

Combining (8.8), (8.9), and (8.10), we obtain the relationship between C/I, frequency reuse plan (N), and path-loss slope (γ) as given in (8.11) and (8.12) for two different models; they are also plotted in Figures 8.9 (a) and (b), respectively.

$$\frac{C}{I} = 10 \ \log\left[\frac{1}{j} \{ \sqrt{3N} \}^{\frac{(44.9 - 6.55)\log(h_b)}{10}} \right] \qquad \text{Okumura-Hata model} \qquad (8.11)$$

$$\frac{C}{I} = 10 \ \log\left[\frac{1}{j} \{ \sqrt{3N} \}^{\frac{(20 + k_d)}{10}} \right] \qquad \text{Walfisch-Ikegami} \qquad (8.12)$$

We find that C/I depends not only on frequency reuse plan but also on number of interferers, propagation environment, antenna height, antenna location, surrounding

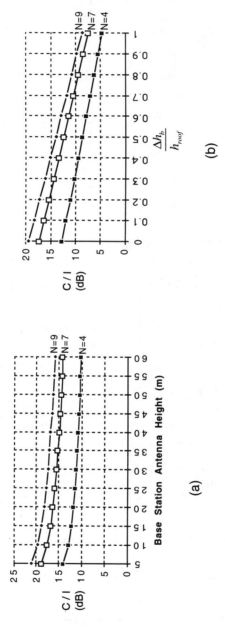

Figure 8.9 C/I as a function of antenna height based on (a) the Okumura-Hata model and (b) the Walfisch-Ikegami model.

buildings, street widths, etc. Antenna directivity also plays an important role in determining the overall C/I performance as we will see in the next section.

8.3 ANTENNA DIRECTIVITY AND FREQUENCY REUSE

Frequency planning optimizes spectrum usage, enhances channel capacity, and reduces interference. A frequency plan also ensures adequate reuse distance to an extent where cochannel interference is acceptable while maintaining a high channel capacity. To accomplish these diverse requirements, a compromise is generally made so that the target C/I performance is acquired without jeopardizing the system capacity. However, the existing frequency planning schemes do not always permit this since the antenna directivity and frequency planning are not coordinated. As a result, cochannel sites are randomly oriented and sometimes points to each other, thus degrading the C/I performance.

In the previous section we noted that an $N < 7$ frequency plan such as the $N = 4$ sectored plan works in dense urban and urban environments but does not work in suburban and rural environments. As a result a hybrid frequency plan (Fig. 8.7) appears to be a good compromise to fulfill both requirements. However, a closer look into these frequency plans (Chapter 7) reveals that the cochannel sites are always randomly directed and there is more than one dominant interferers in each of these schemes. This is due to lack of coordination between frequency planning and antenna engineering. Therefore, a mechanism is needed to coordinate these two design parameters so that the number of dominant interferers is minimized to enhance C/I and capacity.

In this section, we present a method of directional frequency reuse [2] that yields an additional C/I margin and enhances capacity. In this method, a group of channels is reused in the same direction in which the antenna is pointing. As a result, the number of dominant interfering cells is reduced to one, thereby enhancing C/I and capacity.

8.3.1 Principle of Directional Reuse

The proposed frequency reuse plan exploits the antenna directivity to yield additional C/I margin and provide greater frequency reuse, thereby increasing user capacity. In this process, a group of channels is reused in the same direction in which the antenna is pointing. It is based on a tricellular platform comprised of three identical cells, driven from a single source as shown in Figure 8.10. Each cell is treated as a logical OMNI, excited from the corner, separated by 120 deg. Because directional antennas are used in each sector, the worst case cochannel interference is due to only one interfering cell from the same direction. This is shown in Figure 8.10 where the three axes of reuse are subsequently referred to as $\{f(0°)\}$, $\{f(120°)\}$, $\{f(240°)\}$.

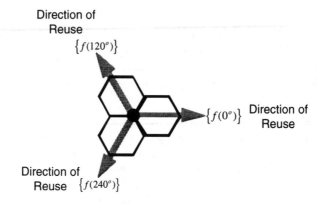

Figure 8.10 A tricellular platform illustrating directional frequency reuse.

Expanding on the principle, we obtain a seven tricellular pattern where each of the three axes is comprised of three parallel layers, as shown in Figure 8.11. These layers are designated as:

$$\begin{matrix} \{f_a(0°)\} \\ \{f_b(0°)\} \end{matrix} \quad \text{along the 0-deg axis} \tag{8.13}$$

$$\begin{matrix} \{f_a(120°)\} \\ \{f_b(120°)\} \end{matrix} \quad \text{along the 120-deg axis} \tag{8.14}$$

$$\begin{matrix} \{f_a(240°)\} \\ \{f_b(240°)\} \end{matrix} \quad \text{along the 240-deg axis} \tag{8.15}$$

The directional frequency assignment results in a total of six or multiples of six frequency groups. Therefore, the available 395 voice channels are divided up into six or multiples of six frequency groups, which are then distributed according to the preceding principle. The principle is illustrated by means of $N = 6$ and $N = 4$ frequency plans in the following sections.

8.3.2 The $N = 6$ Directional Reuse

The $N = 6$ directional reuse plan is based on 18 frequency groups where frequency groups are numbered as $1, 2, \ldots, 18$. These frequency groups are then directionalized and distributed alternately according to the following principle:

$$\begin{matrix} \{f_a(0°)\} = 1, 3, 5, 1, \ldots \\ \{f_b(0°)\} = 2, 4, 6, 2, \ldots \end{matrix} \quad \text{along the 0-deg axis} \tag{8.16}$$

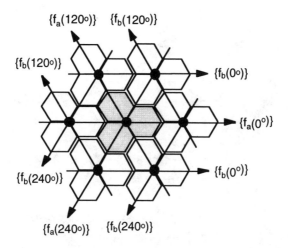

Figure 8.11 Illustration of directional frequency layout.

$$\begin{aligned}\{f_a(120°)\} &= 7, 9, 11, 7, \ldots \\ \{f_b(120°)\} &= 8, 10, 12, 8, \ldots\end{aligned} \quad \text{along the 120-deg axis} \qquad (8.17)$$

$$\begin{aligned}\{f_a(240°)\} &= 13, 15, 17, 13, \ldots \\ \{f_b(240°)\} &= 14, 16, 18, 14, \ldots\end{aligned} \quad \text{along the 240-deg axis} \qquad (8.18)$$

A cluster of seven tricell groups, using the above channel distribution pattern, is shown in Figure 8.12. These reuse patterns are consistently used throughout the geographical service area. The number in each cell represents the frequency group assigned to that cell. As can be seen, frequency reuse within the cluster exploits antenna side-to-ratio, which is exactly 90 deg out of phase. For a typical 105-deg directional antenna, the available side-to-side isolation is >10 dB.

The growth plan, shown in Figure 8.13, exhibits two different reuse distances: $D_1/R = 5.6$, which is due to a reuse site that is 180 deg out of phase, and $D_2/R = 2.6$, which is due to two reuse sites located 90 deg out of phase. The C/I due to the dominant interferer, which is pointing directly at the reused site (i.e., at 180 deg out of phase), can be estimated as follows:

$$\frac{C}{I} = 10 \log\left[\left(\frac{D_1}{R}\right)^{\gamma}\right] + \Delta\text{dB (due to downtilt)} \qquad (8.19)$$

With $D_1/R = 5.6$, $\gamma = 4$, and $\Delta\text{dB} = 3$ dB, we obtain C/I = 33 dB.

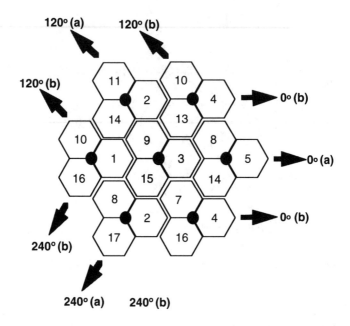

Figure 8.12 Illustration of directional frequency reuse.

The other two contributions from the sides (90 deg out of phase) can be estimated as follows:

$$\frac{C}{I} = 10 \log\left[\frac{1}{2}\left(\frac{D_2}{R}\right)^{\gamma}\right] + \Delta dB \text{ (from side to side)} \qquad (8.20)$$

With $D_2/R = 2.6$, $\gamma = 4$, and $\Delta dB = 10$ dB, we obtain C/I = 23.6 dB, which can be further enhanced by using narrow-beam antennas.

The corresponding theoretical capacity is approximately 395/18 ≈ 22 voice channels per sector and 22 × 3 = 66 voice channels per cell. At 2% GOS, this translates into 14.9 Erlangs per sector and 14.9 × 3 = 44.7 Erlangs per cell.

8.3.3 N = 4 Directional Reuse

The $N = 4$ directional reuse plan is based on 12 frequency groups, providing 395/ 12 ≈ 33 channels per sector. This is accomplished using the same process described in the previous section. The 12 channel groups are directionalized as:

$$\begin{aligned} \{f_a(0°)\} &= 1, 3, 1, \ldots \\ \{f_b(0°)\} &= 2, 4, 2, \ldots \end{aligned} \text{ along the 0-deg axis} \qquad (8.21)$$

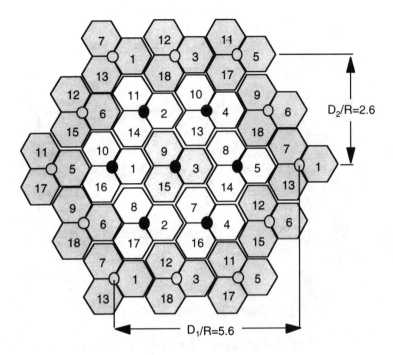

Figure 8.13 Directional frequency reuse growth plan.

$$\{f_a(120°)\} = 5, 7, 5, \ldots$$
$$\{f_b(120°)\} = 6, 8, 6, \ldots \quad \text{along the 120-deg axis} \tag{8.22}$$

$$\{f_a(240°)\} = 9, 11, 9, \ldots$$
$$\{f_b(240°)\} = 10, 12, 10, \ldots \quad \text{along the 240-deg axis} \tag{8.23}$$

Figure 8.14 shows the channel distribution within a seven tricellular platform.

The reuse distances are given by: $D_1/R = 3.732$, which is due to a reuse site that is 180 deg out of phase, and $D_2/R = 2.6$, which is due to two reuse sites located 90 deg out of phase. The corresponding C/I can be estimated as:

$$\frac{C}{I} = 10 \log[(3.732)^4] + 3 \text{ dB (due to downtilt)}$$

$$= 25.87 \text{ dB} \tag{8.24}$$

$$\frac{C}{I} = 10 \log\left[\frac{1}{2}(2.6)^4\right] + 10 \text{ dB (from side to side)}$$

$$= 23.6 \text{ dB} \tag{8.25}$$

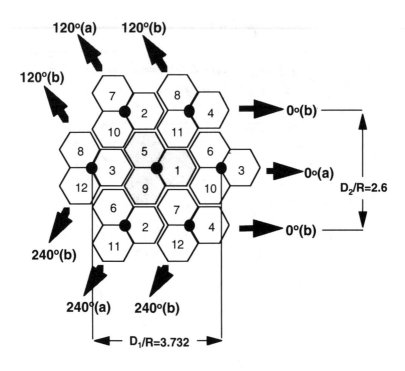

Figure 8.14 $N = 4$ directional frequency reuse.

The channel capacity of this plan is $395/9 = 43.8$ channels per sector and $43.8 \times 3 = 131$ channels per cell. At 2% GOS, this translates into 34.7 Erlangs per sector and $34.7 \times 3 = 104.1$ Erlangs per cell.

8.3.4 Tolerance of Site Location Due to Directional Reuse

Because of an additional C/I margin due to antenna directivity, we modify (8.4) as follows:

$$\frac{C}{I} = 10 \log\left[\frac{1}{j}\{\sqrt{3N}(1 - 2\alpha)\}^{\gamma}\right] + \Delta\frac{C}{I} \text{ (due to antenna directivity)} \qquad \text{for } 0 < \alpha < 1$$

$$(8.26)$$

Figure 8.15 shows the plot of C/I as a function of α. As can be seen, more than 10% relocation tolerance is available by means of a directional frequency reuse plan.

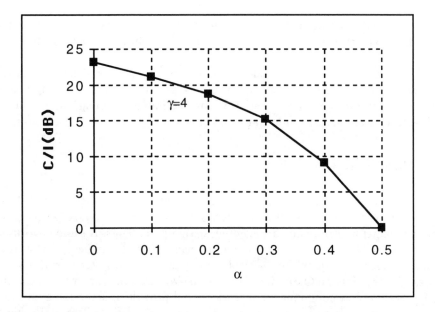

Figure 8.15 C/I as a function of cell site location tolerance for directional frequency reuse.

8.4 PROPAGATION PREDICTION AND LINK BUDGET

8.4.1 Choice of Prediction Model

Propagation prediction is a process of environmental characterization and propagation studies where the received signal level (RSL) is determined as a function of distance. In a multipath environment, the RSL is generally chaotic, owing to numerous RF barriers and scattering phenomena, which vary from one civil structure to another. Building codes also vary from place to place, requiring wide-ranging databases. The computer-aided prediction tools available today generally begin with standard propagation models such as the Okumura-Hata or Walfisch-Ikegami model. These models are based on empirical data and their accuracy depends on several variables such as the following:

- Terrain elevation data;
- Clutter factors (correction factors due to buildings, forests, water, etc.);
- Antenna height, antenna pattern, effective radiated power;
- Traffic distribution pattern;
- Frequency planning;

A detailed description and usage information for computer-aided prediction methods are given in Chapter 7. Table 8.2 provides a quick guideline to the use of these models.

The computer prediction models generally require user-defined clutter factors. As a result, an undesirable error is inevitably present in these prediction tools. Nevertheless, these tools are essential during the initial planning, quotation, and deployment of cellular communication systems.

8.4.2 Model-Based Link Budget

Radio link design is an engineering process in which a hypothetical path loss is derived out of a set of physical parameters such as effective radiated power (ERP), cable loss, antenna gain, and various other design parameters. A sample worksheet is then produced for system planning and dimensioning of radio equipment. It is a routine procedure in today's mobile cellular communication systems. In this section we examine the classical Okumura-Hata and the Walfisch-Ikegami models, currently used in land-mobile communication services, and provide a methodology for radio link design based on these models. It is shown that the link budget is site specific and that there is a unique set of design parameters in a given propagation environment [3] for which the RF link is multipath tolerant.

Table 8.2
Empirical Models

Environmental Zone	Commonly Used Models

Dense Urban
  • Building "canyon" channel propagation
  • Antennas above buildings (macrocell) cause multiple diffractions over buildings
  • Antennas below buildings (microcell) cause diffractions around and reflections on buildings
	Walfisch-Ikegami
	
	Okumura-Hata

Urban • Mixture of various building heights and open areas	Walfisch-Ikegami Okumura-Hata
Suburban • Business and residential areas, open areas, woods	Okumura-Hata

Rural
  • Large open areas
  • Multiple diffractions over obstacles
	Okumura-Hata

These prediction models are based on extensive experimental data and statistical analyses that enable us to compute the received signal level in a given propagation medium. Many commercially available computer-aided prediction tools are also based on these models. The usage and accuracy of these prediction models depend on the propagation environment. For example, the standard Okumura-Hata model generally provides a good approximation in urban and suburban environments where the antenna is placed on the roof of the tallest building as shown in Figure 8.16 to the left. On the other hand, the Walfisch-Ikegami model can be applied to dense urban and urban environments where the antenna height is below the rooftop as shown in the same figure.

These empirical models generally exhibit linear path-loss characteristics, similar to an equation of a straight line (see Chapter 5):

$$L_p \, (\text{dB}) = L_o \, (\text{dB}) + 10 \, \gamma \log(d) \qquad (8.27)$$

where L_o is the intercept and γ is the slope, which depends on the antenna height, antenna location, and the propagation model used. Defining the received signal level as RSL, we can rewrite (8.20) as follows:

$$\begin{aligned} \text{RSL} &= \text{ERP} - L_p \\ &= \text{ERP} - L_o - 10 \, \gamma \log(d) \end{aligned} \qquad (8.28)$$

which can be expressed as

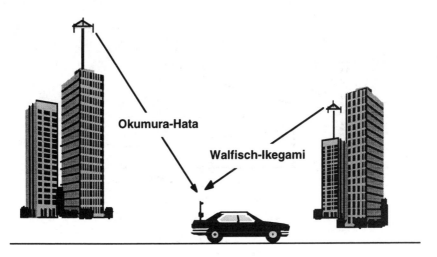

Figure 8.16 Path-loss scenario due antenna location and propagation environment.

$$d = 10^{(\text{ERP} - L_o - \text{RSL})/10\gamma} \tag{8.29}$$

where d = distance; ERP = effective radiated power determined by link budget; L_o = an arbitrary constant determined by the propagation model; γ is the path-loss slope, which also depends on the propagation model, antenna height, and location; and RSL (RSSI in IS-54) = received signal level at the cell edge, i.e., at $d = R$ where R is the cell radius.

Defining $E = \text{ERP} - L_o - \text{RSL}$, we find that (8.22) exhibits several operating conditions:

1. $E = 0$ for which $d = 1$ and is independent of γ (multipath tolerance).
2. $E > 0$ for which $d < 1$ and is inversely proportional to γ (multipath attenuation).
3. $E < 0$ for which $d > 1$ and is proportional to γ (multipath gain or waveguide effect).

These operating conditions are illustrated in Figure 8.17.

It follows that there is a unique set of link parameters such as ERP, RSL, and L_o for which the path loss is linear and independent of γ. In other words, the link budget is site specific and depends on the appropriate propagation model. The ERP for the link budget is therefore

$$\text{ERP} = L_o + \text{RSL} + 10 \ \gamma \log(d) \tag{8.30}$$

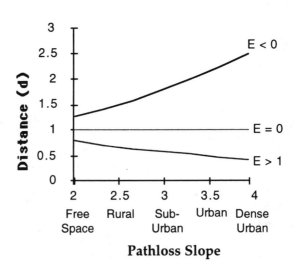

Figure 8.17 Relative coverage as a function of attenuation slope.

where RSL is the received signal level at the cell boundary, which is user defined; d is the average cell radius; and L_o and γ are derived from the propagation model. It is interesting to note that the ERP also exhibits an equation of a straight line with $(L_o + \text{RSL})$ as the intercept and γ as the slope. In terms of the propagation model (see Chapter 5), the link budget then becomes:

$$\text{ERP} = L_o(\text{Hata}) + \text{RSL} + \{44.9 - 6.55 \log(h_b)\}\log(d) \quad \text{Okumura-Hata} \tag{8.31}$$

$$\text{ERP} = L_o(\text{Ikegami}) + \text{RSL} + (20 + k_d)\log(d) \quad \text{Walfisch-Ikegami} \tag{8.32}$$

where

$$k_d = 18 - 15 \, (\Delta h_b / h_{\text{roof}}) \tag{8.33}$$

Example

This example shows the usage of the Okumura-Hata model given the following parameters:

Frequency	900 MHz
Propagation environment	Urban
Antenna location	Rooftop of tallest building
Base station antenna height (h_b)	50m
Mobile antenna height (h_m)	1.5m
Cell radius needed	5 km
RSL (@ $d = R$)	−73 dBm

Determine the ERP.

Answer

Based on the above data, the propagation model to be used is the Okumura-Hata model:

$$a(h_m) = \{1.1 \log(f) - 0.7\}h_m - \{1.56 \log(f) - 0.8\} = 0.015882$$

$$L_o \text{ (dB)} = 69.55 + 26.16 \log(f) - 13.82 \log(h_b) - a(h_m) = 123.337 \text{ dB}$$

$$\gamma = [44.9 - 6.55 \log(h_b)]/10 = 3.377$$

$$\text{ERP} = L_o + \text{RSL} + 10 \, \gamma \log(d) = 50.33 \text{ dBm} \tag{8.34}$$

Therefore, the link budget should return an effective radiated power of 50.33 dBm (20.33 dBW).

8.6.3 Intermod Power

The characteristics of intermod power are shown in Figure 8.20 along with the intercept point. The third-order intermod power increases three times faster than the original signal, as determined by the slope. The intercept point for this product is the intersection of the asymptotes of the characteristics of the original signal and the intermod signal. The slope of the third-order transfer characteristics is steeper than the main signal. Similarly, the slope of the fifth-order transfer characteristics is even steeper, which increases five times faster than the original signal.

The amplitude of the third-order intermod product is given by

$$P_3 = 2P_1 + P_2 - 2P_{3i} \text{ (dBm)} \qquad \text{for } 2f_1 + f_2$$
$$P_3 = 2P_2 + P_1 - 2P_{3i} \text{ (dBm)} \qquad \text{for } 2f_2 + f_1 \tag{8.37}$$

P_1 = output power at f_1
P_2 = output power at f_2
P_{3i} = third-order intercept point.

It is generally difficult to predict the power of these components. However, it is relatively simple to measure and quantify these products by means of a spectrum analyzer.

8.6.4 Cellular Intermod Products

A cell site is a multiple-access point where several channels are combined to form a channel group, which is then transmitted by means of the antenna, as shown in

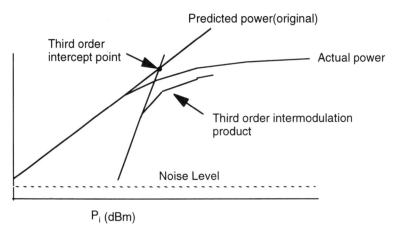

Figure 8.20 Characteristics of intermod power.

Figure 8.21. Intermod products are generated during this process through a nonlinear device such as an amplifier or a corroded connector. These intermod products depend on channel separation within the group, where the channel separation is determined by the frequency plan.

To examine this process, we consider the familiar $N = 7$ frequency plan, based on dividing the available channels into 21 frequency groups, 16 channels per group in the nonexpanded spectrum. Channel separation within this group is given by $21 \times 30 = 630$ kHz. Then a given frequency, say, f_2 can be related to f_1 by means of the following equation:

$$f_2 = f_1 + 0.03 \times 21 \times k \qquad (8.38)$$

where $k = 1, 2, \ldots, 16$ (assuming 16 channels/group) and f_1 and f_2 are in megahertz. Therefore, the third-order intermod products can be written as follows:

$$\text{IM3 (in-band)} = (f_1 - 0.63k), (f_1 + 1.26k) \qquad (8.39)$$
$$\text{IM3 (out-of-band)} = (3f_1 + 0.63k), (3f_1 + 1.26k) \qquad (8.40)$$

and the total number of IM3 products due to 16-channel combinations appears as

$$\text{Total number of IM3 products} = 4 \times 16!/[2!(16 - 2)!] = 2 \times 16 \times 15 = 480$$
$$(8.41)$$

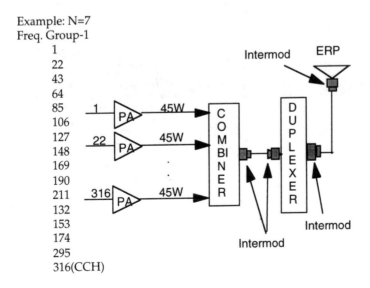

Figure 8.21 Origin of intermod products in a typical cell site.

Translating each channel into the corresponding frequency we find the following:

- In-band IM3 products occupy the entire transmitting band that created them and partially spills over to the adjacent band. See Figure 8.22 for illustrations.
- Out-of-band IM3 products are threefold away from the cellular band and fall within the microwave band.
- The number of IM3 products per 16-channel group is 480.

Similarly, the fifth-order intermod products are given by:

$$\text{IM5 (in-band)} = (f_1 - 1.26k), (f_1 + 1.89k) \tag{8.42}$$

$$\text{IM5 (out-of-band)} = (5f_1 + 1.26k), (5f_1 + 1.89k) \tag{8.43}$$

and the total number of IM5 products due to each 16-channel combination is $4 \times 16!/[2!(16 - 2)!] = 2 \times 16 \times 15 = 480$.

From the preceding analysis we also find that:

- In-band IM5 products occupy the entire transmitting band that created them (Figure 8.23).
- Out-of-band IM5 products are fivefold away from the cellular band and fall within the microwave band.
- The number of IM5 Products per 16-channel group is 480.

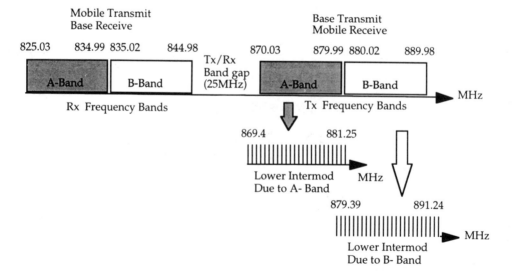

Figure 8.22 Illustration of in-band third-order intermod products in Bands A and B.

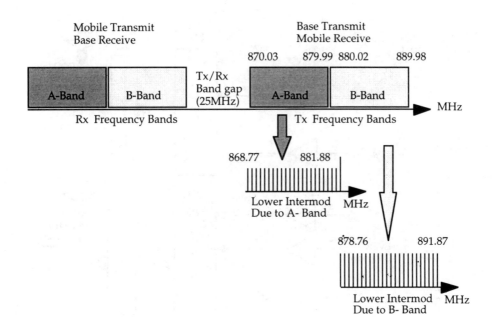

Figure 8.23 Illustration of in-band fifth-order intermod products in Bands A and B.

8.6.5 Intermod Reduction Techniques

Most often, intermod products are generated from the connectors due to high-power transmission. A solution to this problem would be to reduce the power flow though the connectors as shown in Figure 8.24. Here, a 16-channel group is divided into two subgroups of 8 channels each, designated as (1) the odd group and (2) the even group. The odd group is formed by combining 8 odd channels according to the following scheme:

Odd group: Channel 1, Channel 43, . . . , Channel 295

and transmitted through one antenna, designated the odd antenna.

The even group is formed by combining 8 even channels:

Even group: Channel 22, Channel 64, . . . , Channel 316

and transmitted through a second antenna, designated the EVEN antenna. This antenna is generally the diversity antenna, which is normally used for space diversity. The effective power flow, as seen by each path, is now reduced by 50%. As a result, the intermod products are expected to be reduced or virtually eliminated because

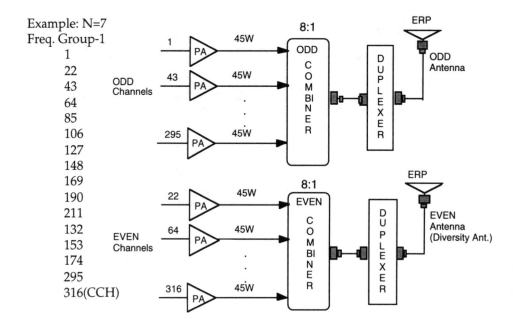

Figure 8.24 Method of reducing intermod based on antenna sharing, which reduces power flow by 50% in each path, reducing intermod problems.

the slope of third-order intermod power is three times the main power and the slope of fifth-order intermod power is five times the main power. Moreover, the effective channel separation, as seen by each combiner is also increased by a factor of 2, that is, $21 \times 2 = 42$ (30 kHz $\times$ 42 = 1260 kHz), reducing combiner insertion loss. Furthermore, the total number of intermod products is also reduced from 480 per group to 112 per group as shown:

Number of IM products in each path = $4 \times 8!/[2!(8 - 2)!] = 2 \times 8 \times 7 = 112$

Figure 8.25 shows an intermod reduced cell site configuration. The transmitting and receiving mechanisms are briefly described next:

Transmitting Mechanism

- ODD and EVEN channel groups are transmitted through separate antennas.
- Both antennas are used as transmitting antennas.
- Channel separation as seen by each path is 1260 kHz. Hence, the insertion loss and intermod products are reduced, enhancing performance.

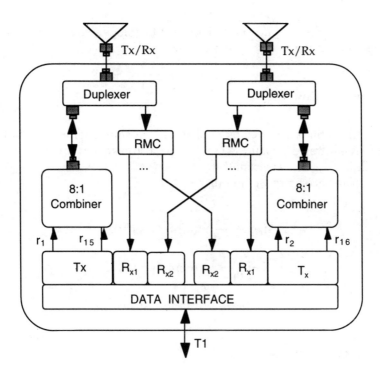

Figure 8.25 Intermod reduced cell site configuration.

Receiving Mechanism

- Both antennas are mutually shared for diversity signals.
- An additional duplexer is needed to perform this operation.

Additional Measures for Intermod Reduction

- Replace corroded connectors.
- Hermetically seal RF connectors.
- Operate the power amplifier in the linear region.
- Increase the dynamic range of the power amplifier.
- Use a high-gain antenna to compensate ERP.

8.7 ANTENNA DOWNTILT

Antenna downtilt is an effective means of controlling the RF footprint within a service area. Two distinct approaches are available to acquire this: (1) *mechanical downtilt* and (2) *electrical downtilt*. These methodologies and their attributes are briefly described in the following subsections.

8.7.1 Mechanical Downtilt

Mechanical downtilt is achieved by mechanically downtilting the antenna from the tower. This is illustrated in Figure 8.26 where we also examine the effects of mechanical downtilt in a cellular environment.

From plane geometry we find that

$$d_1 = \frac{H}{\tan \theta_1} \quad \text{(off-bore sight)}$$

$$d_2 = \frac{H}{\tan \theta_2} \quad \text{(bore sight)} \tag{8.44}$$

$$d_3 = \frac{H}{\tan \theta_3} \quad \text{(off-bore sight)}$$

The respective received signal strengths are:

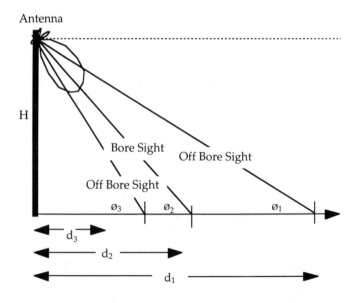

Figure 8.26 Illustration of mechanical downtilt.

$$\text{RSSI}(d_1) = 10 \log(\text{ERP}) - P_{L1} - \Delta\text{RSSI (dB)} \qquad \text{(off-bore sight)}$$
$$\text{RSSI}(d_2) = 10 \log(\text{ERP}) - P_{L2} - 0 \text{ dB} \qquad \text{(bore sight)} \qquad (8.45)$$
$$\text{RSSI}(d_3) = 10 \log(\text{ERP}) - P_{L3} - \Delta\text{RSSI (dB)} \qquad \text{(off-bore sight)}$$

where ΔRSSI (dB) off the bore sight is due to downtilt and depends on the amount of downtilt and vertical width of the radiation pattern.

Although downtilt is an effective way to control the RF footprint, too much downtilt has a detrimental effect on RF coverage. This is shown in Figure 8.27 where poor coverage in the bore sight occurs due to excessive downtilt. In conclusion, three factors, namely, (1) vertical beam width, (2) angle of downtilt, and (3) antenna height must be taken into consideration in this exercise.

8.7.2 Mechanical Downtilt in a Reuse Plan

Mechanical downtilt in a directional frequency reuse plan acquires a greater C/I margin as illustrated in Figure 8.28. This advantage is due to the fact that a reused site resides exactly 180 deg out of phase. This advantage also depends on vertical beam width as indicated in the same figure.

8.7.3 Electrical Downtilt and Greater Frequency Reuse

Electrical downtilt is based on controlling the phase and magnitude in each array element of the antenna. This is generally acquired by means of digital signal processing techniques. Because cochannel sites are stationary, a desired "null" can be produced in those directions only, thus greatly reducing C/I.

Perhaps $N = 1$ frequency reuse can be acquired as a combination of directional frequency reuse and directional electrical downtilt—*food for thought.*

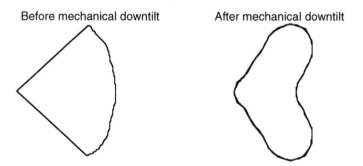

Before mechanical downtilt   After mechanical downtilt

Figure 8.27 Illustration of mechanical downtilt.

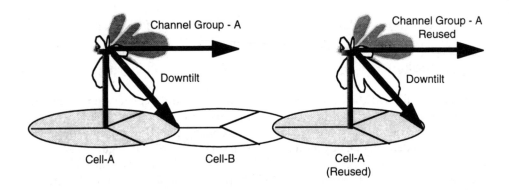

Figure 8.28 Illustration of mechanical downtilt.

References

[1] IS-54, "Dual-Mode Mobile Station-Base Station Compatibility Standard," PN-2215, Electronic Industries Association Engineering Department, December 1989.

[2] Faruque, Saleh, "Directional Frequency Assignment in a Cellular Radio System," U.S. Patent Pending.

[3] Faruque, Saleh, "Propagation Prediction based on Environmental Classification and Fuzzy Logic Approximation," *Proc. IEEE ICC '96*, pp. 272–276.

▼▼▼

ABOUT THE AUTHOR

Saleh Faruque received B.Sc. in physics and M.Sc. in applied physics from Dhaka University, Dhaka, Bangladesh in 1969 and 1970, respectively. He received M.A.Sc. and Ph.D degrees, both in electrical engineering, from the University of Waterloo, Waterloo, Ontario, Canada, in 1976 and 1980, respectively.

At present, Dr. Faruque is an advisor/manager on wireless communication for Northern Telecom Wireless Systems, Richardson, Texas, where he lends RF and system engineering support to AMPS, TDMA, CDMA, and PCS systems. Furthermore, Dr. Faruque is actively involved in formulating and presenting training courses. Northern Telecom has standardized one of his courses, which is now being offered to new recruits and customers worldwide.

Before moving to the United States, Dr. Faruque was working at Northern Telecom in Brampton, Ontario, Canada, where he contributed extensively in the area of wireless communication systems and actively participated in the deployment of Northern Telecom's Cellular products in the Canadian marketplace.

Prior to joining Northern Telecom, Dr. Faruque taught in various universities in Europe, Canada, and Bangladesh. He has published over 48 technical articles in IEEE and IEE journals and conference proceedings. He also chaired a session in the 1992 IEEE International Conference on selected topics in wireless communications.

INDEX

The Artech House Telecommunications Library

Vinton G. Cerf, Series Editor

For further information on these and other Artech House titles, contact:

Artech House
685 Canton Street
Norwood, MA 02062
781-769-9750
Fax: 781-769-6334
Telex: 951-659
email: artech@artech-house.com

Artech House
Portland House, Stag Place
London SW1E 5XA England
+44 (0) 171-973-8077
Fax: +44 (0) 171-630-0166
Telex: 951-659
email: artech-uk@artech-house.com

WWW: http://www.artech-house.com

DATE DUE

MAY 5 2004			